Dear Mary

Thank you so much for your hospitality during the Souter family reunion.

Love
Grant, Deirdre
& Caitlin

Pottersfield Press

The Horses of Sable Island

Barbara J. Christie

Postscript by Zoe Lucas

Pottersfield Press, Lawrencetown Beach,
Nova Scotia
1995

Copyright 1995 Pottersfield Press

All rights reserved. No part of this book may be reproduced or transmitted in any form or by any means, electronic or mechanical, including photocopying, or by any information storage or retrieval system, without permission in writing from the publisher.

Canadian Cataloguing in Publication Data

Christie, Barbara J.

The horses of Sable Island
    Includes bibliographical references.
    ISBN 0-919001-92-0

1. Horses — Nova Scotia — Sable Island — History.
2. Sable Island (N.S.) — History. I. Title.
SF284.C3C47 1995   636.1'009716'99   C95-950042-1

Pottersfield Press
R.R. 2 Porters Lake
N.S. B0J 2S0

Printed and bound in Canada

Published with the assistance of the Nova Scotia Department of Education, the Canada Council and Department of Canadian Heritage.

Front and back cover photographs by Zoe Lucas.

Table of Contents

Acknowledgements		4
Introduction		7
The Horse "Jolly"		9
Chapter 1	The New World	11
Chapter 2	Comings and Goings	19
Chapter 3	The Schooner *Kitty*	26
Chapter 4	They Trot, Paddle, Rack and Waltz	30
Chapter 5	A Son of Trustee	35
Chapter 6	Years of Success with R.J. Boutilier	44
Chapter 7	The Old Patch Mare	49
Chapter 8	Attar of Roses	53
Chapter 9	A Natural Order	61
Chapter 10	Golden Horses of the Queen	63
Chapter 11	"Inured to the Spring of an English Hunter"	70
Chapter 12	Significant dates	81
Chapter 13	I Beg to Report	83
Chapter 14	Strawberries Ungathered	91
Postscript by *Zoe Lucas*		97
Glossary		106
Bibliography		109

Acknowledgements

Although many people have assisted me with my research and it would be impossible to name them all, particular acknowledgement must be made to J. L. Martin, Director, Education Resource Service, Department of Education, Province of Nova Scotia, for his unfailing interest and support, and to Charlotte Myhre, Secretary to the Director of Education Resource Service, for the same.

Many of the staff members of Nova Scotia Museum have been of considerable assistance and my thanks go to: Alex Wilson, Curator of Botany; Barry Wright, Curator of Zoology; Scott Robson, Curator of Buildings and Furnishings and Librarian Susan Whiteside. For invaluable information, special thanks go to Niels Jannasch, Chief Curator Marine History, Province of Nova Scotia, and to Dennis Pulley and Graham McBride, Curatorial Assistants, Marine History.

The Public Archives of Nova Scotia have all but become my second home and staff members have made my research an interesting and enjoyable experience. I am much indebted to Dr. Phyllis Blakeley, Associate Archivist for the Province of Nova Scotia, who has freely given of her wide knowledge. Also of the Public Archives of Nova Scotia staff, I wish to thank Allan Dunlop, Senior Archivist, Dr. Brian Cuthbertson, Public Record Archivist, and Gary Shutlak, Map Archivist, whose help has been constant over a long period of time.

The Public Archives of Prince Edward Island and the Provincial Archives of New Brunswick have given me ready assistance, as has Dr. Charles Armour, University Archivist, Dalhousie University. I am equally indebted to R. J. Graham, Superintendent Station Operations, Department of Atmospherics and Environment Service, and Captain G. J. Williams, District Manager, Coast Guard Base, Department of Transport.

Grateful acknowledgements are made to: Shirley Elliott, Legislative Librarian, Province of Nova Scotia; to Ellen Webster, Co-ordinator, Reference Services Halifax City Regional Library; the Canadian Library of Agriculture, Ottawa, Keeneland Library, Keeneland Race-Course, Kentucky; and to P. Schroven, Seeretary General, Société Royale, Le Cheval Trait Belge for his valuable assistanee in locating Bayard II and Reine, and Dr. A. R. Lock, Wildlife Service.

Much important information has been given me by Daphne Machin Goodall in England and Margaret Gardiner, Kennebec Morgan Horse Farm, Maine.

The helpful discussions held with Dr. D. Welsh were of great assistance, as are his later papers. Terrence Punch has been of continuing help with his wide knowledge and his natural Irish love of horses. Thanks also to Nancy Corston for translations from the French; to Ruth Tobin, R. S. Bouteilier and C. K. Blakeney, who, with their close ties with Sable Island, gave me much of value and interest.

My sincere thanks to Margaret McCurdy who typed my untidy manuscript and with a delightfully light hand on the reins, brought all to completion.

Lastly, I wish to acknowledge Kris Smith and her Sable Island horse 'Flash'. My husband and I visited both some time ago, the horse coming across the field to meet us with autumn sunlight tipping the long hairs of his heavy winter coat. It was about then, listening to his proud owner and watching her grand little horse, that the idea of writing a history of Sable Island horses first came to me. To those two, and many other interesting and helpful people, my sincere thanks.

Barbara J. Christie
Halifax, N.S.
October 1980

For Philip, John and Margot

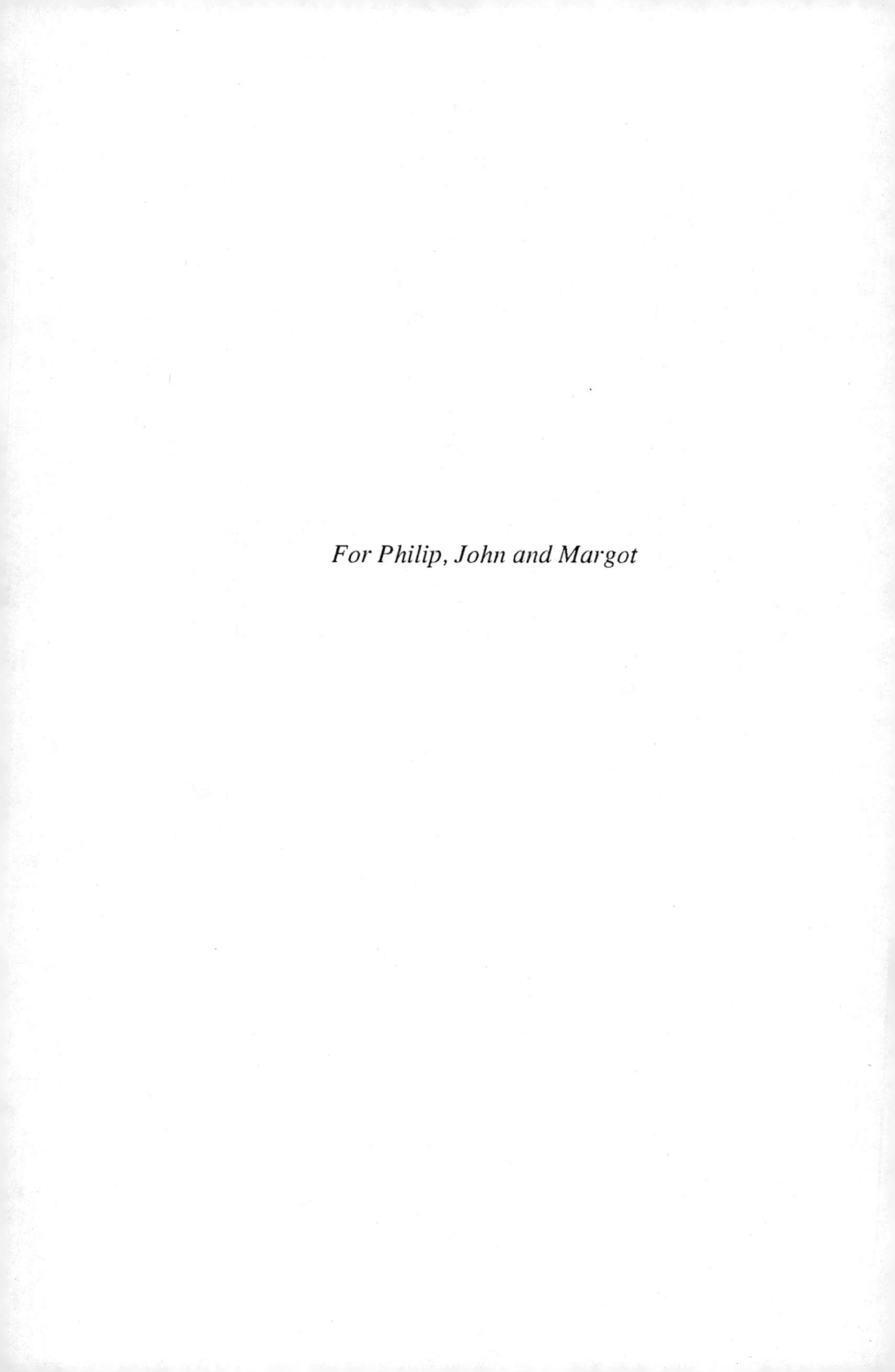

Introduction

Sable Island, with its great running ramparts of storm-driven sand, lies some 100 miles off the east coast of Nova Scotia. It is an enormous sand bank, approximately 20 miles long, covered in large areas with low shrub, grass and, near the shore, wild pea. Converging currents of different temperatures subject it to a good deal of fog while devastating storms constantly vary its basic crescent form. Its treacherous, shallow approaches and terminal bars reach out for many miles and, like the tentacles of an octopus, they have taken down to the depths many hundreds of ships, particularly during the Age of Sail. With justification it has been called the Graveyard of the Atlantic and by some the Dark Island.

It is a sea-bound world of contradiction, and one with unique flora and fauna. Winter can leave it a wilderness of parched growth, at times half buried by the driving sand or snow; summer brings to life the unexpected flowers that grow in the peaty topsoil of the sheltered meadows. Blue iris and bog orchids enjoy the wet land, waterlilies thrive in the shallow pools of fresh water, pearly immortalles, yarrow, golden rod and the sweet smelling wild rose join juniper and other low shrub to compete with the tall golden American beach grass and beach pea for space or sun. The lush seasonal growth of cranberries, crowberries and wild strawberries varied the diet or provided extra income when harvested by the Islanders in the days when Sable enjoyed a population of forty.

It is the home or seasonal nesting ground of a number of birds, both sea and land. Unique among these is the Ipswich sparrow which arrives on Sable Island to breed.

Herds of walrus once made annual calls, but these were all destroyed for their teeth, hides and oil. All that now remains of these sea dwellers are their fellow travellers, the Harbour and Grey seals. They still protest loudly when visitors invade their beaches or, in groups of bobbing heads, keep a sharp eye on things from the near surf. These seals, along with the world famous horses, are the largest animals now inhabiting the island.

This is a brief study of the horses and of a number of additions to the feral herd. Often referred to as ponies, they are by their conformation and history more correctly horses. However, where any such study is made, one horse in particular must lead the parade; a later arrival, yet he stands alone and unique as the first horse there of whom something of his origin is known. His name comes down to us over the years as a light touch from a stern world. They called him Jolly.

Jolly stuck in the quicksands. Drawing and text by the first superintendent, James Morris. (Facsimiles of Crown copyright records in the Public Record Office appear by permission of the Controller of H.M. Stationary Office, London CO 217/76 p.214)

The Horse 'Jolly'

In the fall of 1801 a bay horse stood in his stable at Halifax, awaiting shipment aboard the schooner *Kitty* to Sable Island.

Standing about 14.2 h.h., with his strong build topped off by a fine cresty neck, he showed much of the conformation of the Sable Island horse of today, and was of a type of horse quite common among the mixed breeds of Nova Scotia—the old Acadian horse.

Unless late gelded, the high crest seemed to point to his entire state, and although purchased for general use under saddle and in harness, he would also have been the right type of animal to improve the native island horses. Horses at that time were frequently worked, and also used as breeding or covering stallions.

Named "Jolly", he was to prove an intelligent, courageous animal, and, although of undistinguished breeding, he stands today unique in the island's history. Many horses had been sent there before him, many more were to follow; yet he is the first identifiable horse ever to reach those sandy shores.

Purchased from the owner of the Golden Ball Inn, Halifax, Edward Phelon, he cost the newly formed Sable Island Commissioners, £18, plus the cost of his saddle and harness. He survived many hair-raising adventures in his new home and served his master with faithful honesty for a good many years. He is last listed in the 1812 Sable Island annual inventory and was at that time, eighteen years old.

Chapter 1
The New World

History does not tell us who first ran his boat through the roaring surf about Sable Island or, surviving shipwreck and the crushing weight of the sea, first lay exhausted upon those lonely shores.

We will never know if that first arrival found temporary refuge from the sea, or a sandy prison from which there was no escape. We do know, however, that the magnificent Viking round ships, known as Knarr, beat their way across the Atlantic about the year 1000, and that later, Portuguese, Breton, Basque, English and Spanish vessels came to the fishing grounds off Newfoundland and Acadia (now Nova Scotia) in the early 1500s.

Sable Island could well have been known to these adventurous people who, apparently unlettered, left no trace or record of their annual voyages. It fell to the larger expeditions of explorations, sent out by Spain, Portugal, England, France and the Netherlands, to make the first known records of sightings and landings.

Whatever fortunes may have befallen the island prior to those voyages, the first mention of animals there, either native or introduced is that listed in the accounts of the ill-fated voyage of Sir Humphrey Gilbert in 1583. While his fleet anchored at St. John's, Newfoundland, a Portuguese sailor told them that, "Some thirty years past they [the Portuguese] did put into the Island both neat and swine which were multiplied exceedingly." Animals of the bovine genus were at that time termed "neat" cattle. Some historians and story-tellers have assumed that horses were released at the same time, and therefore reason that the modern Sable Island horses descend from them. However, let us look at the possibilities.

When on their three to four-month voyages, the Portuguese fishermen fed on pork or beef, biscuits, wine, olive oil and vinegar. It is known that some animals were set out on islands as a source for fresh meat or for the use of shipwreck survivors. All fishermen had mainland bases where they salted down or dried their catches, but to have put animals ashore there would have been to lose them to Indians or wild animals.

There is a lack of reliable documentation as to these early

comings and goings so that facts and fable become a colourful mix, however, there may be some connection between the story of Sir Humphrey Gilbert and that of the Portuguese animals on Sable Island.

It is documented that in 1521 the King of Portugal granted Joào Alvares Fagundes a large territory embracing what is now Nova Scotia and adjacent land, together with various islands lying off shore. These included the island of Santa Cruz (Sable Island). Fagundes is said to have sent out several vessels for exploration or settlement. One group of three lost two vessels off Sable Island. What could be salvaged was loaded aboard a small caravel for settlement elsewhere; the animals—cows, goats, sheep and pigs—were abandoned on the island.

Some cloven hoof prints have been found on Sable Island at a depth that could date back to the Fagundes animals but nowhere in the black and peaty depths of the island has been found the imprint of an equine hoof; and nowhere is found reliable documentation of their being there until hundreds of years later.

In 1627 the Company of New France had been formed to increase crown revenues from the North American fur trade and to bring order to settlement. Isaac de Razilly, cousin of the powerful Cardinal Richelieu who in 1624 had created the first breeding program for horse improvement, was designated lieutenant-general of all parts of New France, called Canada, and Governor of Acadia. On September 8, 1632, de Razilly arrived at La Hève, Acadia, with his vessel escorting two transports carrying 300 souls, farm livestock, seeds, tools and agricultural implements. The first continuing agricultural development in Acadia stemmed from this expedition and later shipments. Included with these were two other important arrivals: the first French peasant women and what is all but certain in the light of later research, the arrival of some horses. With de Razilly's connection with the Cardinal and the importance of his new command it is more than probable that he would have brought out horses of an improved type for service there.

All future Acadian horses descend from these, as yet undocumented, equines of the 1631-35 period, not from those which had arrived earlier at Port-Royal in 1612. These with their new foals were removed in 1613 when Captain Argall, from Jamestown, Virginia, destroyed the settlement and removed all livestock.

La Hève was soon to have an interesting visitor in the person of John Rose of Boston, Massachusetts. Rose had sailed out of Boston in 1633, and a short time later, his vessel, the *Mary and*

Jane, was wrecked on Sable Island. He was there for several months building himself a small yawl from the wreckage, and during this time noted about 800 head of cattle grazing there. He reported these to the Acadians when he managed to sail to La Hève. Under pressure, he acted as a guide, and returned to the island with the French, who built temporary houses and commenced rounding up the cattle to slaughter, mainly for their hides. Such was the devastation that by the time Rose finally returned to Boston with the news of his find, hastily formed New England expeditions to Sable Island found the herds all but destroyed. By 1635, only 140 head were said to remain, a very few animals to chase over 40 odd miles of loose sand.

Marc Lescarbot, lawyer-historian at Port Royal, Acadia, in 1606, wrote of an earlier and unsuccessful attempt at settlement by the French Baron de Lery who left various animals at Canso and on Sable Island. Lescarbot wrote one hundred years after the supposed event, and his account is given little credence by serious historians. There is all but no evidence to support the existence of the Baron's expedition and by one authority it has been termed "utterly improbable".

The ill-fated expedition of the Breton nobleman Troilus de La Roche de Mesgouez (Marquis de La Roche-Mesgouez) reached Sable Island in 1598, on the way to attempt colonisation on the mainland. Those left on the island were a very poor type to establish settlement, being some 50-60 convicts.

Few meat animals were landed with them, and although the colony was supported by at least two supply ships from France, long periods of neglect, plus the quarrelsome nature of the men, reduced their number to eleven within five years. No horses were recorded by the survivors and, but for the Portuguese animals already there, all the men would have perished.

Another theory which has been considered is that horses swimming ashore from Spanish shipwrecks were the tap root of the modern herd. Although storms at sea can force vessels hundreds of miles off course, the general area of Spanish exploration and settlement lay far to the south. It seems unlikely that Spain would have lost anything but fishing vessels at such northern latitudes. There is no evidence that either Spanish or Portuguese fishermen put horses on the island.

The "horses from the sea" theory crops up repeatedly in horse history, and there is no doubt that animals have made their way at times from wrecks. However, in weakened state after long periods at sea, few would have survived the battering of storm, shipwreck and the difficult struggle to land. Horses from the defeated Spanish Armada of 1588 are said to have influenced the breeds of horses in England, Scotland and Ireland, but this is doubtful. The great Spanish galleons, storm-driven and fleeing from English pursuit, ran short of water in the North Sea long before rounding the Shetlands in their retreat to Spain. At that point the fleet commander, Medina Sidonia, ordered all horses and mules thrown overboard. Hundreds of dead and still swimming animals were seen, but so far out at sea that their hope of reaching the Scottish, or any other shore was virtually nil. Long before the remaining vessels passed near Ireland, all expendable horses had been jettisoned. This was often the fate of animals at sea when a vessel encountered difficulties.

The blood of imported Barb, Turk, Spanish and Arabian horses was employed in designed breeding to improve many types of breeds of horses in Great Britain, Ireland and the European continent. Because these fast and often fiery horses were so valued, their influence has often been claimed for horses of totally unknown origin in order to increase the price of prestige of the animal. For those who see Barb blood in the conformation of some of the Sable Island horses, and lay undue emphasis on it as a clue to the origin of these animals, it would be wise to pause and make a study of horses in early Acadia, New England and Canada, before reaching a conclusion.

By 1634, Sable Island, along with Port Royal and La Hève, had been granted by Isaac de Razilly to his brother Claude. Claude de Razilly was mainly concerned with the fisheries and played his part in the removal of what was left of the Portuguese cattle, personally profiting by their destruction. In 1635, Isaac de Razilly died and left all his rights and property in Acadia to his brother Claude, who conveyed the whole to Charles de Menou d'Aulnay, in 1642. D'Aulnay's struggle for power with his fellow countryman, Charles de La Tour, became his prime concern. Sable Island, stripped of its cattle, was of no further importance to either man and was abandoned.

Nicholas Denys was concerned with the fisheries, timber trade and was involved with various posts and positions in Acadia until 1669. He controlled the territory of Cape Breton, the shores of the Gulf of St. Lawrence westward to Gaspé, and probably the Ile Saint-Jean (later, Prince Edward Island). In his valuable survey, *Description and Natural History of the Coasts of North America*, (p.207), he writes of Sable Island: "There is no longer anything upon it except the pond and some grass, there being no cattle left. These having been killed solely to obtain their skins". Contemporary accounts of the island agree with Denys, and yet there appears a puzzling contradiction.

In his 1894 address, entitled "Sable Island: Its History and Phenomena", to the Royal Society of Canada, the Rev. George Patterson, D.D., included in a printed Appendix A, the following: "Of the destruction of these cattle by the Acadians, we have another notice in a letter by Bishop Saint Vallier, written in 1686, after a visit to Acadia."

After describing Beaubassin, the bishop reports,

> About ten years ago the first French came to this place from Port Royal. In the beginning they were obliged to live chiefly on herbs. At the present they are in more easy circumstances, and there is an abundance of pasturage in the vicinity, they have let loose a number of cows and other animals, which they brought from Sable Island, where the late Commandant de Razilly formerly left them. They had become almost wild, and could only be approached with difficulty but they are becoming tame little by little, and are of great advantage to each family, who can have a good number of them.

These records show that from the 140 head of cattle said to have been on Sable Island in 1635 by Claude de Razilly, stock was removed up until 1642. It seems hardly likely that he would have turned around and restocked the herds as has also been said was done. But some may have remained and it is possible that Denys, who did not visit Sable Island personally, was misinformed. The island was much larger then, so some animals could have escaped notice. Even today, with the advantages of motor transportation and aerial photography, precise counts of the horses are difficult to make.

The cattle of the Portuguese were red in colour, yet the Acadian cows are said to have been black. There is one other possible origin for the Beaubassin cattle. The Acadians engaged in a good deal of illicit trade with New Englanders and, thrifty though the French were, it is possible that it was from the English that the cattle were obtained by the common barter process. Perhaps the good bishop was led a little astray with the harmless story of the ready Sable Island supply of farm animals.

Soon after 1641, New Englanders drove cattle through to the Swedish settlements on the Delaware, supplementing the small number of animals brought out by the colonists. Certainly by

1699, it was from these same English that Acadians were obtaining stallions and breeding mares.

If there were still cattle on Sable Island, one wonders why a French report of 1697 reads thus:

> Two leagues from Pechmoucady [Passamaquoddy] is an island called Grand Manane, which is fit for feeding cattle only. If one were to put some bulls and cows there, and forbid anyone to hunt on the island, in less than five or six years the King could collect enough Salted beef for whole armies at the sole cost of its transportation to France. Monseigneur, Your Honour must be aware of the quantity of such beef that the English unseasonably collected from the island of Sable.

This was decidedly wishful thinking, considering the temper of the times.

New plans for Sable Island were made in 1738, when the Rev. Andrew Le Mercier, a Huguenot minister of Boston, petitioned Nova Scotia's Governor in Council for permission to rent land on the island. (Acadia was then under English rule and the French name of Acadia had been changed to that of New Scotland, or, Nova Scotia.) Le Mercier first applied in March of 1737 and the application was considered the following year. The Reverend gentleman had interesting associates in the proposed venture, namely the powerful Boston merchant Thomas Hancock, Henry Aitkins and, later, John Gorham.

The listed aims of the petitioner include:

> what great Advantage it would be to the Publick in General and particularly to His Majesty's Unfortunate Subjects and to all others who may at any time prove so unlucky as to Suffer Shipwreck near to or upon the island of Sables, was that Island well settled with proper habitants: and Stocked and furnished with Cattle of all sorts.

Not waiting for permission to rent (which permission seems never to have been granted), the associates, "with these good intentions," sent out, in 1737, "Horn Cattle, Swine, Sheep and so forth. In Order to Succour, Help and Relieve such as may be shipwrecked there."

Whatever his original intention, Le Mercier did not settle his own family on the island; but working crews and some families did farm the land there for fifteen years. The seal hunt was carried on, and lives were saved when ships were wrecked. However, the endeavour was not of a purely humanitarian nature. Thomas Hancock was heavily engaged with West Indies trade; animals and produce raised by the settlers could be periodically shipped off to the mainland and sold or, on being transhipped, sold in the islands to the south.

History shows that little in the affairs of Sable Island over the years has been of a purely selfless nature; too often there has been exploitation, pillage, and savage slaughter of the animals by raiders, to say nothing of some pretty open wrecking from time to time. Le Mercier was no more grasping than the next man in those rough and tumble days of early North American settlement, but that Huguenot divine did concern himself with turning a nimble penny from time to time and on one occasion involved himself in not a few problems when he attempted to sell church property to his own advantage.

The periodic cattle raids continued and at Le Mercier's complaint, Nova Scotia attempted to halt these by passing stiff laws for the island's protection. But when Le Mercier's associates fell away and he was unable to find navigators willing, or capable, of making voyages there and back to Boston, he again advertised his island holdings for sale, with absolutely no legal grounds for doing so. In *The Boston Weekly News Letter* of February 8, 1753, he gives a comprehensive account of his holdings and their ability to produce a profit. He adds,

> When I took possession of the island, there were no four-footed Creatures upon it but a few Foxes, some red and some black, now there are I suppose about 90 sheep, between 20 or 30 Horses, including Colts, Stallions and breeding Mares, about 30 or 40 cows, tame and wild and 40 Hogs.

There is some confirmation of this story in the account of the loss of the Irish snow, *Catherine*, on the east bar of the island on July 17, 1737. Carried in the same newspaper for August 11, of that year, the news item goes into considerable details of the wreck, her cargo and the passengers who survived. It reports the making of shelters and the repairing of one of the vessel's boats in which some sailed to Canso to seek help. Taken off on July 25, they had eight days in which to make some exploration of their sandy haven — something which can be done in good weather from the high dunes and hills. Although it was a common occurrence for survivors to report the animals or people found there, these survivors made no mention of either. Perhaps, as Le Mercier said, there were none, and his had not yet arrived.

In 1737 Le Mercier makes the same observation as Nicholas Denys had in 1671, that there were no large animals on the island. What is more important is this first recorded date for horses arriving on Sable Island. A study of the modern horses there, still running free, must commence from this date, and not that of the earlier introduction of horned cattle by the Portuguese.

Chapter 2
Comings and Goings

We do not know what became of Le Mercier's horses after his people vacated Sable Island, or if he was successful in selling his holdings. The inventory of his estate after his death in 1764, makes no mention of the island, or of any possessions he may have retained there. His heir, Andrew Le Mercier, Jr., appears not to have inherited his father's rights, and some indications exist that these may have been handed over to Thomas Hancock to meet debts incurred during their partnership.

Both John Gorham and Thomas Hancock continued to be interested in the island. Gorham unsuccessfully applied for private land grants in 1744. Hancock seems to have felt his rights were unchallenged and "Some time before 1760, fitted out a schooner and sent to Sable Island horses, cows, sheep, goats and hogs." As with the earlier animals sent there during the partnership, it was said to have been for the benefit of shipwreck survivors. Regardless of the doubtful single intent in the introductions, Hancock's animals were to be of marked importance in the history of the Sable Island horses, and circumstances strongly suggest that their original owners had been the Acadian French.

Had Le Mercier been unsuccessful in finding the purchaser for his island holdings, it is probable that he would have at least removed some of the animals in order to close the venture with something to his financial advantage. His advertisement of 1744, illustrates well what would have happened to the remaining animals. This long notice offers a reward for information as to the identity of those responsible for raiding and pillaging of the settlement. In frustration he wrote,

> Not withstanding these two [Nova Scotian] Proclamations, the love of Money, which is the Root of all Evil, is so deeply rooted in the Hearts of some Fishermen, that they have sundry times stole our Cattle and our Goods, regarding neither the Laws of God or Man; neither Justice to me or Humanity to shipwreck'd Men, which by their Wickedness they endeavour to starve, and minding neither natural or revealed Religion and their Eternal Damnation, nor even their own temporal interest, which is certainly not to hinder,

but to promote, the said settlement since it may be their case one Time or other to be cast away upon the said Island Sables, and to want there, those things which they have carried off.

Such thieving raids were commonplace and continued on, well into the nineteenth century, when even the strategically sited lifesaving huts, with their supplies of food and warm clothing, were looted. It can well be imagined that between 1753 and the time Hancock's cattle arrived that the island would once again have been all but stripped of livestock.

The futility of restocking the island with fresh animals for whatever purpose, only to have them stolen or killed, might in normal times (for that period in history) have stayed Hancock's shipment. However, great changes were in the air and with his keen business sense, the merchant would have seen personal gain in the changing fortunes of others. He had long been involved in the affairs of Nova Scotia. In the 1749 founding of Halifax, he had been official local agent of supply for building materials and food; during the attack on the French fort of Beausejour, his contribution to the success of the action had been some twenty thousand pound sterling in cash and war materials; in November of 1755, he and his partner, Charles Apthorp, supplied shipping on a large scale for the transportation of a very sorry cargo; human beings caught in the power struggles of England and France.

By the 1713 Peace of Utrecht, the Acadians, unconquered by war, were nevertheless transferred to English authority by agreement with the Imperial French government. Fortified Port Royal became Annapolis and the province's name was changed from Acadia to Nova Scotia. A period of one year was fixed during which those Acadians not wishing to take the Oath of Allegiance to the English Queen Anne, might sell their land and leave the province with all their livestock and moveable possessions. After that period they were free to leave but would then forfeit everything but what could be carried. The twelve months brought agreement from some, but for the most part, a many-sided struggle began that was to continue for forty years: the governors in Halifax coping with conflicting pressures from England and those from the colonists in New England; the Acadians torn by a basic wish for independence, but counselled to resist English rule in the name of their uncaring Motherland by their local advisors. The situation after the long years of struggle was much like that of two women at odds in the same kitchen, one must leave before peaceful operations can begin. The decision to expel the Acadians was at last arrived at in the fall of 1755, and for the main body of those tragic people, swiftly executed.

Although periodic deportations were to continue until 1762, those leaving in 1755 numbered some 6,000 of an estimated total of between 11,000 and 12,500 souls. The numbers of animals abandoned at that time, again, can only be estimated, but figures arrived at by Andrew Hill Clark in his book entitled *Acadia, The Geography of Early Nova Scotia to 1760* are a pretty sound guide. For the years 1748 to 1750, the estimates of livestock at Annapolis, Minas Proper, Pisiquid (now Windsor) Cobequid, Chegnecto and other outlying settlements, these figures are given: cattle—17,750, sheep—26,650, swine—12,750, horses—1,600. The numbers would have increased by 1755, and although not all of these would have been abandoned at that time, the figures show that a very great number of animals would have been affected by the total expulsion.

A thousand or so head of the abandoned animals were driven to Halifax, another drive took horses and cattle through the woods to Lunenburg, some may have gone by sea. Others were corralled at Windsor where hundreds were slaughtered and the meat salted down. The demand for salt was so great that supplies ran out in the province.

Six of the better horses were taken to Windsor at the request of the Governor, Colonel Charles Lawrence. This was not necessarily for his own use, but as a common practise in times of war so that good breeding animals could later be put to best use. It is said that the greater number of horses were given to the soldiers involved in the general operation. Many of these would no doubt have changed hands to the nearest buyer for what in those days bought a great deal--a gallon of rum; but where was the local market for them, and to whom could they sell in the depopulated province, other than to traders such as Thomas Hancock? Such merchants could ship them out to other markets or arrange for their care and later sale in the province.

As well as his involvement in the removal of the Acadians, Hancock was also made Nova Scotia's Boston agent for New Englanders willing to settle on the vacated French lands, and here again he could profit. Not all of those people would have wished to risk losing their livestock on a possible rough sea passage; it would be easier to sell their animals in Boston and to purchase Acadian cattle at a good price on arrival at their new homes. Outside of the city, Thomas Hancock just happened to have extensive grazing lands, areas which, by common right, belonged to the people of Boston. What better place to hold cattle until required for shipment to the West Indies? It is all but impossible to believe that Hancock did not profitably involve himself in the Acadian cattle bonanza and in the disposal of any cattle available, prior to the departure of the New England Planters. The probability that the animals put on Sable Island by the merchant were of Acadian origin is strengthened by the following facts. The precise year when Hancock fitted out his schooner and sent animals to the island is far from certain, although "sometime before 1760" was a guess made many years later. The new stock was taken out at some period during the years of the Acadian expulsions; from November of 1755 to 1762. Thomas Hancock was dead by 1764, and his Boston shipping records do not list such a shipment to the island for the seven years. If animals were available in New England and in Nova Scotia, certainly in such numbers as could be sent as store cattle to an island, wouldn't any sound businessman ship them by the safest, shortest and cheapest means?

The distance from Boston to Sable Island is 550 nautical sea miles; four-and-one-half days for a first-class vessel and one week for a lesser craft. The distance from Chebucto Head, outside of Halifax Harbour, to the island is 147 nautical miles, 18 hours sailing for a schooner of the type used by Thomas Hancock. In both cases, due to the treacherous approaches, 14 hours of the trip would have been under soundings, in order to avoid shipwreck on the island's surrounding shoals.

To connect the modern Sable Island horses with those of the Acadians and to compare similarities in both, particularly with regard to the often noted Barb characteristics, something of the latter must be considered. Little is known of the early Acadian horses as to their origin, type and numbers. Andrew Hill Clark writes: "there seems to be almost a conspiracy to hide the presence of horses from us." Acadian records are not plentiful, and over the years, they and later historians have made little note of horses.

We do know that in the 1680s the Acadian equines were of fine build, that they had good shoulders, clean, strong legs, lasting hooves and that their heads were a little large. As with the horses of Quebec, then called Canada, they are said to have been of Breton and Norman breeding; two of the most serviceable breeds for light draft and saddle use. Robert Leslie Jones deals at length with the Quebec horses in his "The Old French-Canadian

Horse" (*Canadian Historical Review*, Vol 28, No. 2) and uses Henry W. Herbert's description.

> The Canadian is generally low-sized, rarely exceeding fifteen hands and oftener falling short of it ... His characteristics are a broad, open forehead; ears somewhat wide apart and not infrequently a basin [concave] face; the latter perhaps, a trace of the far remote Spanish blood, said to exist in his veins ... His crest is lofty, and his demeanour proud and courageous. His breast is full and broad; his shoulders strong, though somewhat inclined to be heavy; his back is broad, and his croup round, fleshy and muscular. His ribs are not, however, so much arched, nor so well closed up as his general shape and build would lead one to expect. His legs and feet are admirable; and bone large and flat and the sinews big, and nervous as steel springs. His feet seem almost unconscious of disease. His fetlocks are shaggy, his mane voluminous and massive, not seldom, if untrained, falling on both sides of his neck, and his tail abundant, both having a peculiar crimpled wave ... He is extremely hardy, will thrive on anything, or almost nothing; is docile, though high spirited, remarkably sure-footed on the worst ground, and has a fine, high action, bending his knee roundly and setting his foot squarely on the ground ... He is said, although small himself in stature, to have the unusual quality of breeding up in size with larger loftier mares than himself, and to give the foals his own vigor, pluck and iron constitution, with the frame and general aspect of their dam. This by the way, appears to be characteristic of the Barb blood above all others, and a strong corroboration of the legend, which attributes to him an early Andalusian strain.

Although the horses of Canada and Acadia were of the same breeding, those of Acadia arrived earlier and came, one might almost say, via private enterprise; those in Canada under stricter government control. The blood of each group eventually became influenced by the importation of mixed breeds from America, but from 1665 until 1671 those of Canada were improved exclusively by breeding animals sent out from France.

In 1665, Jean Baptist Colbert, Louis XIV's brilliant minister, reestablished the National Stud, or Haras, earlier created by Cardinal Richelieu. In his capacity as secretary of state for the *maison du roi* and for the navy, as well as the controller general of finance, Colbert's influence was considerable. As Canada was subject to the French Court of Admiralty, the multi-hatted minister would no doubt have been aware of the two stallions and twelve mares first sent to Quebec in 1665. Periodic shipments continued until 1670, and Montcalm described them as being much like the horses of the French Ardennes. Although, no doubt somewhat changed, the same breed of horse was used as "wheelers" (team nearest the gun carriage) by the French artillery during the 1914-1918 war. Colbert continued his interest in the shipments of horses, and wrote in 1671, "I shall see that mares and she donkeys are sent to Canada, in order to encourage the increase of these species so necessary to the comfort of the settlers." The ministers may have seen the citizens of New France concerning themselves with these lowly animals, but the horse-loving *habitants* had other ideas. By 1710, the rapidly rising numbers of horses in Canada was alarming the French authorities: too many of the settlers' sons were putting all their ambition into raising fine saddle horses. Fears were expressed that the Canadians would lose their famous skill as walkers, or no longer make use of snow shoes, thus losing their superiority over the English.

Interbreeding of the Canadian horses and those of the English colonies began after 1671, when Jean Talon, the great Intendant of New France, decided no further expensive importations were required from Colbert's Haras. Future needs for breeding stock could be filled by purchasing horses from the growing English colonies to the south. Although well informed as to the affairs of the Acadians and a great admirer of their sheep, Colbert does not appear to have favoured these colonists with shipments of breeding horses—he probably knew his Acadians. Long before Talon looked to the English for breeding stock, these independent minded descendants of de Razilly's settlers filled their need for new blood by turning to their old, but forbidden trading partners, the New Englanders. By the late 1600s, Acadia's Governor, Joseph Robinau de Villbon, was regretting such traffic, but recognized "... our need to obtain from them stallions and mares for breeding purposes, so that the stock may be completely changed." The horses on whose influence Villbon's fears were based were the descendants of earlier horses brought to the New World by Spanish, English, Dutch, Finnish and Swedish colonists. Several shipments of Irish horses had also arrived in Virginia during 1620. Degrees of Spanish, Barb or Arabian blood could be found in the English, Irish, Dutch and possibly the Swedish horses.

One notable example was the Chickasaw horse, standing some 13.2 h.h. bred by the Choctaw, Cherokee, Seminole and Cree Indians, which represented this rich blood in the largest measure. They were in fact, naturalised Andalusians; small descendants of a larger type of animal, first brought across the Atlantic by the early Spanish colonists and explorers. The Virginians bred these fast, tough horses, or ponies, to their superior animals and produced an early type of short-distance race horse; the forerunner of the modern Quarter horse.

While a shortage of horses in the English colonies did exist early on, by the time the Acadians would have been buying from the New England traders, a fair range of choices were available. Virginia had a shortage in 1649, but Massachusetts and Rhode Island had a surplus of horses by the following year. By 1668, the Virginians not only caught up with the other two areas, they were beginning to have problems with the ever increasing numbers of their equines. Bands of near wild horses, many the offspring of the highly bred animals of the colony's cavalier element, careened through the streets at night or damaged crops. The irate and sleepless Virginians thinned their ranks by sale and by export. Here then was a plentiful supply of horses—animals that could be picked up for a song and moved on to the Acadians at a nice profit.

Although various breeds and types were freely interbred in the English colonies, it is probable that the Acadians imported stallions and mares from there that would ensure continued stamina, size and quality in their own breed. As the Canadian French were to do at a later date, they may have looked for a type of horse that carried a high degree of Barb, Spanish or Arabian blood, thus adding still more of these rich bloods to that already in the Acadian horse. Little wonder then, that the Sable Island horse, should he be of Acadian origin, exhibits some Barb characteristics. They could come not from the romantic, mythical wreck of a Spanish galleon, but from the rather more mundane interbreeding of the New World's horses. It is a characteristic that has not been bred out by the influence of other stallions sent out to the island; introductions which were to begin arriving soon after settlement was established there by Nova Scotia in 1801, and who, in several known stallions, further increased the influence of the Barb.

Chapter 3
The Schooner *Kitty*

When Thomas Hancock died in 1764, Sable Island lost a controlling hand, at least as far as a business interest in the animals was concerned. Wrecks increased in numbers and despite the efforts of Nova Scotia to prevent such depredations, fishermen and casual opportunists helped themselves to salvage and whatever else the island could provide. Another decade brought even greater problems for the cattle and horses when both factions in the American War of Independence extended their battlegrounds to Sable Island. Each removed or slaughtered cattle for beef until all were gone; horses were taken for remounts. General havoc was wreaked so that nothing there would be of service to either side.

By the time hostilities ceased, island lawlessness had become the norm, prompting observations such as this in 1800,

> So Cruel and unfeeling of late years have been the number of wretches, as have taken Cargoes of those horses from the Island, and carried them for sale to the West Indies. [for use in the sugar mills and plantations]. Many having been wantonly shot by persons Wintering on the Island for the purpose of Wrecking and their skins brought to this Town [Halifax] and other parts of the Province.

Shipwreck survivors were stranded at all seasons of the year with little food and less shelter until sighted by passing vessels. Monies accruing from the sale of horses, hides and tallow, as well as salvaged cargoes and much from the wrecks, were falling into the hands of near criminals instead of provincial coffers.

Sir John Wentworth, Nova Scotia's lieutenant-governor at the time, like several of his predecessors, had long been aware of the need for permanent island settlement. Previous marginal efforts to establish this had failed; England was too far away and had more pressing concerns. Neither financial help, support nor encouragement of any kind had been forthcoming through her various secretaries of state. Sable Island had been defined as "an Appendage of Dependancy of the Province of Nova Scotia." Through the efforts of governor, provincial assembly and council, decided to go ahead alone in creating a humane establishment.

Plans were made and commissioners were appointed from the provincial council and assembly.

One shipwreck in particular had brought the need for the expensive project into clear focus. This was the loss of the *Frances* on the night of December 2, 1799. She had been carrying members of the staff and household of His Royal Highness, Edward, Duke of Kent, the newly appointed Commander-in-Chief of His Majesty's North American Forces. Also on board were many of his valuable possessions and supplies, together with twelve beautiful horses and those in charge of them, a coachman and four young stable boys.

There were less illustrious, but still wasteful and tragic losses. An item in a Halifax newspaper of June 11, 1801, reads:

> We are sorry to learn that a large ship of 300 tons, from Boston, bound for Liverpool, England is cast away on Sable Island. She went ashore in the night, with all sails set. Her cargo is valued at £26,000 Sterling ... part of her cargo is said to be saved, and was there a proper establishment on the Island it is thought the whole might be preserved.

The newly formed Sable Island Commission began to move.

James Morris, with a fourteen-year career in the Royal Navy behind him, was appointed the island's first superintendent. Plans were drawn up, frames for a house, barn and storehouse fabricated, and stores for at least six months began to be stockpiled. He would be required to

> carefully preserve and diligently to encourage the growth and increase of all cattle, Horses and other livestock, which you will find upon the Island or are sent there by the Commissioners... And you are not to suffer any to be exported, upon any account or pretence whatever without Licence from this Government first hand and obtained ... report quality of stock and whatever measures may be useful for its preservation and increase. If any is absolutely necessary to be used for comfort and recovery of shipwrecked persons, or other sick persons, let it be done with the greatest discretion and frugality and the same reported to the commissioners at Halifax.

James Morris came of an old and talented family and the men going out with him were experienced and able. There was no mistaking the purpose of the operation: Sable Island was no longer to be a Tom Tiddler's Ground where all might do as they pleased.

The schooner *Kitty*, 47 tons burden, built at La Have the year before, was chartered to carry supplies to the island. Her master, William Crook, had taken her to Spanish River (now Sydney) for sea coals. Then, after a thorough cleansing, James Ives had pens and a platform built on her for the new settlement's animals.

While the horse, Jolly, waited in his stable, other animals for the use of the settlers began arriving at dockside; most of them from the Windsor and Falmouth areas. With the horse the last on board, the *Kitty* must have looked like a nineteenth-century Noah's Ark, for besides house frames and timber, her full complement comprised four men, "One bull, two in-calf cows, two in-pig sows, one boar, one male goat, one in-kid goat, eight ewes, two rams, nine ducks, ten dunghill fowls." It is recorded that all of the animals were of the very best kind.

Escorting the *Kitty* was His Majesty's armed brigantine, *Earl of Moira*, commanded by Captain Fawson, carrying James Morris, his wife and family, together with additional needs and supplies. The two vessels took a total of fourteen people to new homes; more than half of the children. The schooner cleared Halifax, October 5, 1801; the brigantine sailed the following day. Morris stood on deck as they passed and hail the *Kitty*. Reading his journal today, one can still feel the pride and quiet excitement he felt as he embarked on his new commission. They were sentiments enjoyed by many future superintendents. The island was for them, in many respects, a small kingdom, or a titanic ship permanently anchored at sea.

After a rough passage, both vessels arrived at their destination October 10. Equally rough and difficult landings were made on the next three days— "The stock much bruised and fatigued." However difficult the operation had been, they had been spared a storm that could have wrecked the venture while it was still at sea. On the return voyage to Halifax, the *Earl of Moira* encountered severe gales that sprung her bowsprit and main mast, "the Brig," reported Fawson, "straining much in her upper works that I was obliged to throw over the lee guns and cut away the boat from the stern." As many Captains before and certainly after him, Fawson must have felt a sense of relief as he approached the outer reaches of the harbour with his mission safe and successfully accomplished.

With winter fast approaching, the island settlers made haste to erect the house and store shed. Until that was accomplished, tents were their only shelter. However, Morris's troubles were just beginning. "But surely," he wrote the commissioners, "the carpenter who framed the house must have been in love or stupid, as many pieces were wrong numbered and no braces of any consequence ... some pieces missing which gave me a deal of trouble." Timber from wrecks had to be found and shaped to fit. During the delay, rain soaked much of their provisions before it could be got under cover. Bread (hardtack) salvaged from another hulk sustained them. Morris thanked a merciful God for the windfall, "Without it [our] sufferings would have been considerable."

By November 6, the buildings were in place and a flagstaff had been erected. A rough lean-to type of stable, with sloping thatched roof, had been built into the side of a sand hill. Jolly and his bovine companion found their first haven from wind and cold. In typical make-do island fashion, it also became a maternity home, when the first sow to farrow hauled her pregnant bulk up the hill and produced her piglets in the deep straw of the thatching.

There were to be continuous improvisations before Morris could inspect the native horses. The carpenter was not the only one to be stricken with love sickness when lists of needs for the settlers were compiled. No butter churn had been included and Mrs. Morris had to make do with the barrel of a cannon and a cartridge case. Running short of fodder, seals were shot to provide food for the hogs. Jolly was too light for the job of hauling heavy timbers from the shore and soon found himself double-yoked to the island's bull; neither taking at all kindly to the ignominious hitch and uneven pull. Finally a day arrived when both horse and master were free from their labours and together began the first of many journeys of exploration. The Sable Island horses most probably all descendants of those left there by Hancock had some surprises in store for each.

Chapter 4
They Trot, Paddle, Rack and Waltz

Anyone who has ever kept goats, particularly a buck goat (male), will read Morris's report of November 1801 with sympathetic understanding.

> All the animals remain domestic, the Goat too much so. I am sure no ape or monkey is more antic. I am obliged often (tho with reluctance) to beg his life. Sometimes he breaks peoples shinns, butts the dogs, poultry and my neighbour's door has often been drove down by him. He has killed two ducks and once pushed the girls in the fire ... I expect we must banish them to the east end.

He is referring, of course to the goats, not his daughters!

Whether the banishment took place or not we do not know. However, that end of the island was soon to be inspected and wrecks, livestock and general conditions noted. The initial encounter with the horses was to be something of a jolting experience for both Morris and Jolly.

> One of the horses, more bold than the rest, came for me full speed, blowing his nostrils and making a dreadful appearance. I was in no great alarm for my own safety, but the horse I expected to be destroyed or terribly mangled ... the animal came within about four yards and made a stand ... he was dark bay, about 15 h.h. a mane below his breast, and stamping his foot like a sheep at a dog.

Morris had yet to learn how the Sable Island stallions challenge intruders, particularly a strange male horse. Although somewhat alarmed both horses and Morris came through the encounter unscathed; one other stallion was not so fortunate. Members of a shipwreck crew were out about the island on a duck hunt. Charged by a master stallion and, like Morris, not knowing that the stallions will charge, but rarely attack humans, the animal was shot. Morris illustrated and wrote an account of the incident. The marksman had only one eye, but with confidence in his musket, stood his ground and fired at close range. The ball entered the neck fifteen inches from the breast bringing the animal down, but it immediately got to its feet and ran for half a mile before expiring.

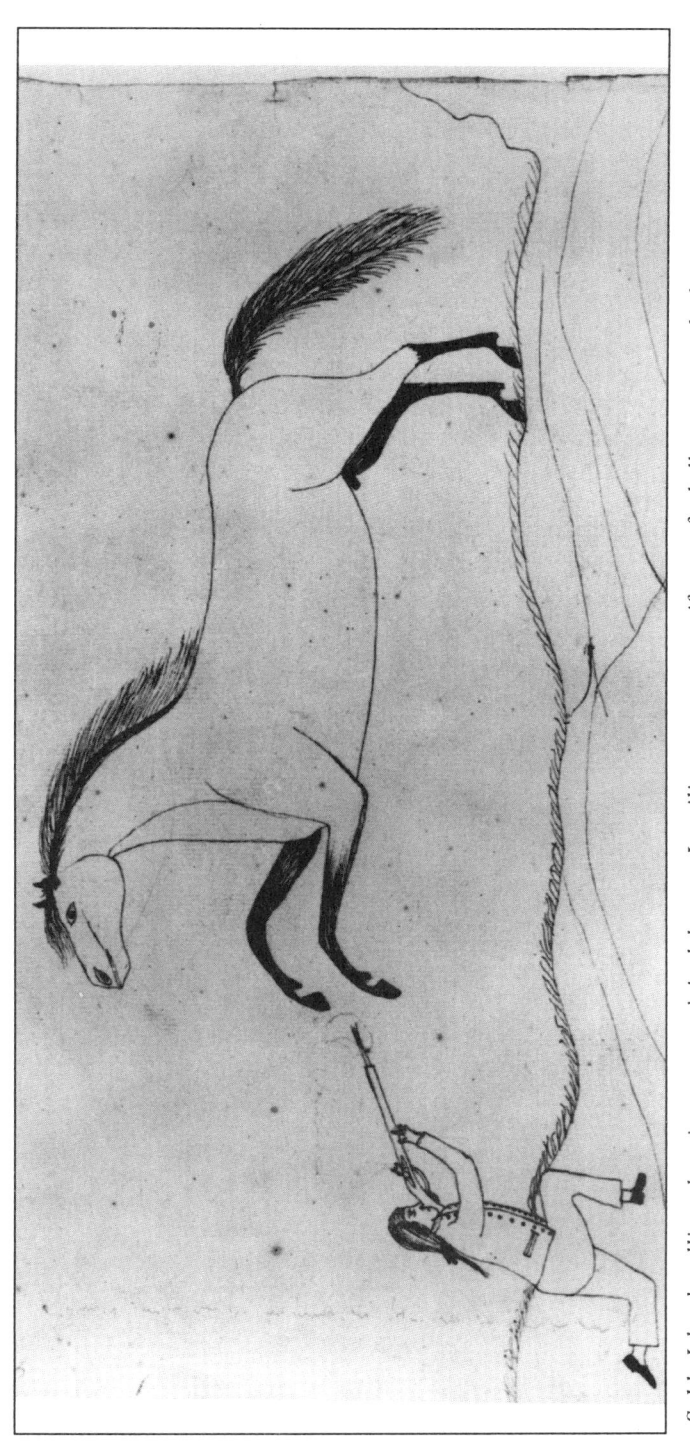

Sable Island stallion charging one-eyed duck hunter. In military manner, as if part of a hollow square, the hunter goes down on one knee and, with good eye to his site shoots from the left shoulder. (Facsimiles of Crown copyright records in the Public Record Office appear by permission of the Controller of H.M. Stationary Office, London CO 217/76 p.206)

Other than his own close call, Morris seemed pleased with his first sight of the horses, taking note of several small herds totalling sixty-one head. "They are in general bay coloured, a few black and only two iron greys and two pied white and dark bay [skewbald]."

His overall assessment of them after several rides of inspection in 1802, gives us a better picture:

> The Wild horses of the Isle are in general a middling size. Thick, short neck, thick legs and I expect three quarters of them are bay—dark and light—and the others various coloured, some reddish black pied [Skewbald-brown and white, Piebald-black and white]—whitish and grey—very few of the latter ... They are very wild and fleet, have in general a very handsome trot and canter. I expect there is about ninety on the island ... Their principle haunts [as today] are at and about the East end of the Isle ...

By the time these first rides were over Jolly must have wished himself safely back on the mainland. Morris camped out on the long trips, tethering his mount close by. Ending one day tired and cold, Jolly developed colic and despite efforts to alleviate the animal's distress, the horse continued to roll in agony. Morris went to his tent that night despairing for the survival of his mount and expecting to have to face a fourteen-mile hike alone the next day. The tough little horse astonished him the following morning; not only was he on his feet again, but had wakened his master by calling loudly for his breakfast.

By one other illustrated account, we know that Jolly came close to losing his life shortly after when he became stuck in quicksands. Morris managed to get himself clear, but felt Jolly was "irretrievably lost". But, "groaning like a human in distress," the horse managed to get a footing on some submerged wreckage and dragged himself onto firm ground.

Adventures such as these with their animals and other members of the island community, often bound them all together in a close association of a nature that few mainlanders experience. Even in retirement, they were homesick for the island, and children born there were an exclusive band who could meet years after leaving the island and relive with enjoyment their unique childhoods.

Morris, however, was the first of his kind. While he was breaking new ground on Sable Island, the commissioners in Halifax were in very much the same position. There were no well-defined guidelines and his naval training did not prepare him for livestock management, certainly not that on Sable Island, with its special needs and conditions. Although at first enchanted with his new world, the stress of rescue work, salvaging, caring for wreck survivors, plus general garden, crop and farm management dampened his enthusiasm. He wrote nostalgically of missing the "sea-coal smoke" of Halifax and then there was his constant tussle with the horses.

How the first to be landed at Halifax were gathered in, he did not say, but certainly those that followed gave the islanders a rough and exhausting time of it. No one knew just how to go about capturing them, which is difficult to believe when one considers the numbers sent off or stolen in past years. It is recorded that old illustrations of the 1700s showed them being captured by mounted men with lassoes, a method used about the same time for the capture of the free-ranging mountain ponies in Wales. Morris tried "snares" and driving without success. Those that they managed to capture broke their bonds or pens. A ditch was dug, but the horses, driven down the sloping ramp, turned and scrambled out again, leaving only a mare and her young foal. As he had noted earlier, the horses were wild and fleet and very much at home in the deep island sand that was to lame or ruin so many riding horses in future drives. Jolly could not be spared for what was frequently, up hill and down gully, a horse chase for 20 or so miles. Later the native horses were broken and used for this purpose and their incredible ability to keep their feet was noted by all who rode them. Adding to Morris's earlier comments on their gaits and action, an 1885 observer writes,

> The horses trot, jump, gallop, paddle, rack, prance, shuffle, and waltz ... They will carry a stranger anywhere without the slightest inconvenience to themselves ... and in the wild round-ups ... need no urging whatever to race uphill and down, turn, double head round, take a flying fifteen feet leap over a gulch and drive back any who make a desperate struggle for their freedom ... The chase over ... the jaded little Tartars desired no stable or groom, but as soon as their saddles and bridles were removed, they started off for a roll in the grass, a drink at the pond, and the wild freedom of the [sandhills] untill their services were again required.

These big drives were in the future. Morris did not have the advantage of a "motley cavalcade" of herders to assist him, each "equipped according to fancy; red shirts and blue, rough pea-jackets and stained tarpaulins, hats and caps of fantastic shape, and flaunting bandanas wound around the head, all mingled in a curious melange, bobbing as they go, like corks upon the waves." Early captives were taken for island use; the first four sent off in 1803 found favour with the commissioners, two of them going to Sir John Wentworth and his son, others to local officials. All appeared pleased with their acquisitions and no disparaging comments as to their quality or conformation were made; the systematic removal of better animals that was later to weaken the herds was still in the future. Although the numbers shipped off remained light and the count for 1804 was 150-odd head, the years between then and 1808, saw their numbers greatly reduced. Old mares were killed for hog feed and colts and stallions of all ages were culled to supply meat for the establishment and wreck survivors.

The following year, care of the island was to rest on new shoulders. The dedicated, resourceful but often frustrated Morris, beset by family tragedies, increasing island demands and ill health, died in October shortly after landing from the *Swan* on his return from Halifax. He was replaced by Edward Hodgson, a veteran mariner who had served under Sir John Wentworth and the great cartographer, Col. Joseph Desbarres.

Chapter 5
A Son of Trustee

During Hodgson's superintendency (1809-1830), some 35 vessels were wrecked on the island or its approaches; 1050 lives were saved. Such rescue work nearly always involved hauling lifeboats long distances over the sandy beaches to the wreck position. Following the saving of lives, attempts would always be made to salvage cargo, or material from the wrecks themselves. Any delay often resulted in the seas claiming all within a few short hours. Strong horses were needed for such operations and the light native horses were not up to the long hours of gruelling work, so heavier draft animals began to be sent out by the commissioners.

In 1811, £26, 1 shilling were paid to Jacob Miller, important merchant of Halifax, for an in-foal "draught mare", 12 shillings and 6 pence for a cart collar and 8 shillings and 6 pence for a pair of hames. She arrived safely soon after June 20, on the schooner *Hercules*. The following July of 1812, the schooner *Martha*, brought out another bay mare, also bred, and costing £22, 10 shillings. The first mare died with her unborn, full-term foal the same year, and one other mare sent out in 1813, slipped her foal in passage.

The two remaining mares both produced foals in later years and in 1820, Hodgson advised Halifax, "There are two elegant Mare Colts here from the Mare you sent... which I should be glad to send off if you should wish it." The establishment must have been up to strength with its work animals for these to have been surplus, but the very fact of their existence raises the question as to their sire. Either they were the progeny of Jolly, or the mares had been covered by the Sable Island stallions.

The numbers of horses, both native and establishment bred, sent off to markets in mainland Nova Scotia, New England and Newfoundland, increased with the years and it is known that there was a periodic introduction of stallions for the purpose of breed improvement in the island's horses These must have arrived during the period 1801-1830, but all were reported as being killed by the native stallions. The list of those that were to follow is of necessity incomplete.

For a period of 130 years after the 1801 establishment was formed by Nova Scotia, only 54 years are covered with any degree

STRIPPING THE WRECK.

of accuracy as to the introduction of new breeding stock. Some superintendents did not record their arrival, and local reports that may have noted them have been lost, as have many of the journals themselves. Less precise and scattered evidence as to their existence is found in private memoirs, newspapers and in some provincial reports, such as that of 1850, compiled by Nova Scotia's distinguished statesman, Joseph Howe.

Fourteen stallions can be identified with certainty; several are of impeccable breeding and their colour, weight and age are known. Against a background of extensive research, others come partially into focus; an additional six, at least, must remain in the shadows until time and future research gives them substance. Thus, we have a possible total of 20 stallions, and to these must be added 9 brood mares, all identified as to breed, and one with an extended pedigree.

Joseph Darby succeeded Edward Hodgson and remained until 1848. A man of furious energy and unbending conviction, he expected his staff to respond to the Sunday bell calling all to his house for church service, regardless of their exhaustion at times from round-ups or rescue. Inventive, as were so many of the islanders, he constructed a 50-foot portable wharf and a "capstan to heave it up the beach". This facilitated greater ease of loading horses and island cargo into the surf boats that carried them to the larger vessels lying at anchor offshore. Not only was Darby conscientious and efficient at his job, but he cut the American Beach grass for rough fodder, cultivated the 10 acres of garden first established by his predecessors, and bred some of the largest and better quality horses for the early days of island herd management. Despite demands for marketable horses, he attempted to retain good brood mares.

During the latter stage of his superintendency, there was considerable friction between Darby and the Sable Island commissioners and the root cause of this may have lain in their demands upon him as to the management of the island; not at all an uncommon occurrence in Sable Island's history. Regardless of whose design it was, the fact remains that during the last years of his office many of the best of the stock, including brood mares and promising breeding animals, were shipped off. It was ten years or more before the herd recovered from this catastrophic culling.

Darby shipped off a total of 300 horses. The herd numbered 70 head when he arrived. At his departure the count was 250 head. There is no record of any mainland stallions being sent out,

Lifeboat truck and team with crew, c.1890, Sable Island, Nova Scotia. (Courtesy, Miss R. Tobin)

although new blood may well have produced the superior horses regrettably absent in later years.

M. D. McKenna of Shelburne, Nova Scotia, succeeded Joseph Darby, and was on the island from 1848 until 1855. In his journals we find the first of listed stallions. On May 4, 1853, the government schooner *Daring*, brought out the stallion Napoleon. Described in a Sable Island commission report as "young" and "a superior and wellformed animal ...", he fell sadly short of someone's assessment and expectations. Turned loose with a small gang of mares after the native master stallion had been removed, he proved unsatisfactory. Described by McKenna as being "externally deficient as a proper stallion" he thrived on island living, but appeared to lack some of the more vital attributes of his Corsican namesake by "letting any shabby thing come into his gang and make himself at home." Kind and easily managed in the stable, he was returned to Halifax in 1856 to commence a rather more mundane career as a saddle horse. His breeding is unlisted, but he may have been of predominantly Acadian French or Canadian French blood. However, some very good thoroughbreds and a son of Sherman Morgan had been standing in various parts of Nova Scotia for some years for the purpose of breed improvement, but there is rather more indication that he may have been a cross of the first two breeds.

In 1853, 52 Sable Island horses were shipped off in the American schooner *Smith Tuttle*. Her owners, Isiah and Frank Gifford of Provincetown, Massachusetts, sent the big schooner and her master, Captain Rowley, to the island in July, where he paid £113 for the following:

> 5 horses of 1 year old @ 20 shillings
> 22 horses of 2 years old @ 30 shillings
> 15 full grown horses and mares @ 60 shillings
> 10 mares with colts @ 60 shillings

The need for men and horses to work well together is illustrated in McKenna's vivid report of the 1854 wreck of the 715-ton *Arcadia*, out of Antwerp on October 26 of that year. This fine copper-bottomed vessel ran aground in dense fog at six o'clock in the morning.

A strong team to haul the largest lifeboat had to be hitched and horses saddled to carry rescue crews. Then began the long beach journey to the north east bar. McKenna wrote, "the *Arcadia* lay two hundred yards from the beach, settled deep in the sand and listed seaward with her lee side under water, main and mizzen mast gone by the deck and a tremendous sea running and sweeping over her bows." The island crew found that the mate of the *Arcadia* and some of her crew had made it to the bar, but had been unable to return with aid to the vessel.

The island lifeboat was launched and with *Arcadia*'s mate and the island men aboard, a start was made for the wreck

> contending with tremendous seas, strong currents and high winds they got alongside the [*Arcadia*] and during the afternoon made six trips and brought on shore about eighty persons large and small. Two other attempts were made to reach the wreck but the oars and thole pins were broken by the violence of the seas and the boat had to return to the beach. An attempt was made to run a warp from the ship to the shore, but the current ran at such a rate that it could not be accomplished. When night came on and we had to haul up our boat the cries of those left on the wreck were truly heart rending. In the hurry of work families had been separated and when those on shore heard the cries of those on the wreck at seeing the boat hauled a scene was witnessed that may be better imagined but cannot be described. I walked slowly from the place leading my horse till, by the roaring of the sea, the whistling of the winds and the distance I had travelled their doleful cries could not be heard, and then I took my seat in the saddle."

That lonely, agonized walk away from the wreck scene and so much unavoidable suffering, rather than a casual mount and ride for home, that might have given the impression to those aboard of unfeeling desertion, is typical of so many moments shared by the island men and their horses. McKenna had already ridden forty miles that day in the loose island sand; both he and the horse had stood hour after hour on the stormy beach. For the animal there was a warm stable and rest, but the superintendent still had to oversee the feeding and housing of the survivors. The injured had to be cared for and preparations organized for the next day when, fortunately, all were safely taken off the wreck and united with their loved ones. McKenna adds, "The ship was broken in a thousand pieces [by a second storm] on the night of the 29th, and only a few packages of cargo and some small things of ships materials are saved." The need for an immediate and all-out effort by both man and beast when a wreck was sighted is clearly illustrated by this last entry.

During 1855, two more stallions arrived at Main Station; the first on the *Wave*, June 15, and the second on the *Daring*, August 9. One was of unknown breed, but likely part-bred; the other, well known and documented. This latter was the blood-bay Thoroughbred Retriever, his pedigree always given as by the American-imported, but English-bred, Trustee, out of Caprice by Muley, etc. In Nova Scotia, he was first owned by Charles Hill Wallace of Halifax, where he raced mile and mile-and-a-half flat race heats in the annually held races on the North and South Commons between 1841 and 1843. His owner for 1845 was G. G. Gilbert of Saint John, New Brunswick, by whom he was entered in two match races with the stallion Livingstone, also by Trustee, and who was owned by Hugh McMonagle of Sussex Vale, New Brunswick.

The races run on the sands of Courtenay Bay, New Brunswick, saw Retriever win his first match but, out of condition and with a back injury, go down to defeat in his second. Afterwards retired to stud duties, his bills list him as owned by James Peters of Saint John, New Brunswick. By 1850, he was again in Nova Scotia and standing at stud at the Spring Garden Stables, Halifax. His owner at that time was given as Mr. Tobin. There is evidence that he was sold for £50 by John Lithgow, a Halifax wine merchant, to the then supervisory body for Sable Island, the Board of Works for Nova Scotia.

While his son, Young Retriever, out of the imported English bred mare, La Belle, by Pollio, out of Maiden by Hedley, still stood in Nova Scotia, and his stud notices continued in local papers, those of his sire disappeared in 1854, when he left for Sable Island. Not suited by age, condition and years of racing to island conditions, he served for a while to some effect, but was returned to Halifax in 1857. He was 19 years old.

Superintendent McKenna left Sable Island in 1855, and was replaced by Philip S. Dodd, during whose time both Napoleon and Retriever were returned to the mainland. Only one stallion of the usually listed "improved breed" has been found to arrive on the island as a replacement. He arrived in October of 1861 and could have been the son of these stallions imported for mainland breed improvement some years before.

One was the imported English thoroughbred, Sir Rupert (his name was later changed to Norfolk). This horse was the bay son of Sir Hercules, out of Miss Tree by Merlin. Another stallion was the son of a dapple chestnut offspring of Sherman Morgan, his dam had been a Bellfounder mare. He had arrived in Nova Scotia in 1842, and had been bred by Samuel Jacques of Ten Hills Farm, near Boston, and was called Bellfounder Morgan. Also standing was the American bred sorrel thoroughbred, Tornado. This stallion was by Boston, the sire of the incomparable Lexington. Hants and Kings counties had the services of Sherman Morgan, a stallion named after Justine Morgan's son, Sherman. Cape Breton had another Canadian called Napoleon, a black, "hard looking horse, tough as a hemlock knot". Lunenburg and Queen's counties had a chestnut stallion, Sir Henry, a "noble animal, fit for all work". Yarmouth and Shelburne stood Membrino, also known as Thomas Jeffery, "the picture of a carriage horse". Richmond and Inverness had a Green Mountain Morgan, an animal carrying the proud old Justine Morgan blood through his sire, Hale's Green Mountain Morgan by Gifford. Sydney and Guysborough travelled Black Hawk, by Long Island Black Hawk. He is said to have been three parts bred. A Morgan colt called Messenger had gone to Annapolis and Digby counties, and to Halifax and Colchester counties had gone the powerful grey Canadian, crossed Clydesdale, St. Lawrence. Of all these Nova Scotian stallions, a choice of their offspring for stallion duties on Sable Island would probably have fallen, as it so often did, among those with a fair share of Morgan or Canadian blood.

The yearly export of horses from the island continued and Dodd was facing the ever present problem of retaining good breeding stock. He wrote to his superiors in 1857,

> With reference to the shipping of Horses mentioned in your letter, I beg to state, that it would be difficult at present to muster a cargo without shipping many of our young mares, which I think would interfere materially with the improvement of the Breed of the island as most of the Old mares are evidently failing. I had a difficulty last year to collect the few I sent off in consequence of so many being shipped just before I came to the Island.

From 1857 to 1885, the journals of the superintendents and other reliable records are either missing or fail to make note of any new stallions arriving. The daily entries show the usual everyday working of the island staff, and the never ending battle with wind and drifting sand. In wet weather horse slings were made for hoisting the horses from surf boat to supply vessel; harness was repaired and new sets made; nails were manufactured from wreck metals and all lifesaving apparatus was carefully checked. Fine weather in season saw them carting loads of eel grass from the centre lake to work into the sand of the gardens. This grass was also used to bank the sides of the barns and houses during winter and it was routinely used to spread over areas that the wind had begun to erode. Hay making and cranberry harvest would put all hands to work, while hauling timbers from the beach to wood piles or saw-pit was another running chore. Horse round-up and periodic shipwrecks were times of near exhaustion for all, while then, as now, the arrival of supply vessels and visitors made a welcome change from day-to-day island life.

D. McDonald from Stewiacke, who followed Superintendent Dodd and was island "governor", as they were often called, had charge of Sable from 1873 to 1884, with J. H. Garroway acting superintendent for seven months during 1884.

In July of 1879, McDonald had his hands full when the *City of Virginia*, carrying passengers and cattle ran aground and both humans and animals had to be rescued and re-shipped. While dealing with these pressures, McDonald was furious to see a schooner approach the wreck with its name covered by a piece of canvas and only the word "Gloucester" visible. While McDonald watched helplessly, one large upright piano, by Broadways of London, packed in its zinc case, was removed, as were "three arm chairs, covered in red velvet", drapery material, the ship's silver and a trunk full of silver and other possessions belonging to one of the passengers. A reward of $200 was later offered for information leading to the arrest of the "Gloucester fishing pirates". No doubt the ghost of the Rev. Le Mercier would have read the ad with sympathetic understanding.

Chapter 6
Years of Success with R.J. Boutilier

The journals of Robert J. Boutilier, probably the most able and successful of all island "Governors", are widely informative and are all but complete for his term of office. Running from 1884 to 1911, they carefully list many aspects of good herd management and give a vivid account of busy island life.

First taking up his position when just 29 years of age, this native of French Village, Nova Scotia, was ambitious and energetic. Working towards producing superior horses, he had gelded or removed all entires from the island's west end, "except those on which the operation could not be performed in consequence of the breed running out". And in July, 1885, he turned loose, with fourteen native mares, the dark bay, part-bred, Jack of Trumps. Bred by W.H. Church of Old Clays Farm, Pereau, Kings County, Nova Scotia, his sire was the American bred Climax, by imported Balrownie. His dam was the flat racing mare Linnet, by Retriever. Church, Nova Scotia's most noted and successful jockey, owned and raced three of these horses; importing Climax from the United States, and breeding Linnet and Jack of Trumps. The bay stallion Jack of Trumps raced on the old Mystic Racetrack in Boston, at Bangor, Maine and all over the Maritime provinces.

He changed hands a number of times and last raced over hurdles on the Halifax Riding Grounds course in 1884. He was used for a short period as a teaser stallion at the stable of Thomas Robinson of Doyle Street, Halifax. Robinson sold him to the Dominion government (Sable Island was under the control of Fisheries and Marine at that time) in June 1885.

Running free with his mares and covering well, he was to lose his life at the end of his first season. Fifty horses had been sent off to Halifax from the east end in late September and the "old fighter" stallions, short of mares, came west to challenge Jack of Trumps for his. Although the raiders had already travelled fourteen long, sandy miles, they made short work of the "tame" stallion who went down under their collective harrying and attack. Worn out in the defense of his mares and badly wounded in the shoulder, he was driven into the quicksands off the Lake Beach Road, where he died. Boutilier was shaken by the tragedy and tightened his management so that no further introduced stallions were lost in this manner during his control of Sable Island.

In September of that year, 1885, the colt, Telephone, arrived on the Government vessel SS *Newfield*. He was unbroken and almost certainly of trotting-Standardbred-stock. His new custodian enjoyed breaking the latest acquisition and drove him along the wide, sandy shores in a sulky by way of winter exercise for both horse and man. He is last listed as a stock horse there in 1898.

Looking for more weight to throw into the collar, Boutilier's next stallion, Samson, arrived in 1886. Again, his breed is not recorded, but the name and other evidence points to a degree of Clydesdale blood stemming from an earlier Nova Scotian importation from Lanarkshire. His first offspring were shipped off the island as five-year-olds; others supplied the island stations with more suitable animals for the heavy work of salvage and lifeboat haulage. Samson is last listed in the year 1894.

The 1892 arrival had an interesting background and sire. In 1847, J. C. Wilson of Maine, U.S.A., purchased a chestnut-roan stallion while travelling the banks of the St. Lawrence River in Quebec. This 14.2 h.h. horse, whom he named Flying Frenchman, was reputed to have been sired by a "spotted Arabian", who was said to have been salvaged from a river wreck. Here again, we have a horse who suddenly appears from wreck and water. Wilson's new purchase had raised, almost black spots on darker patches on his quarters. These, together with his indomitable spirit, fire and pacing gait (all to be found in the Canadian horse) he was to pass on to many of his progeny. Those less impressed by the wreck theory considered him to be of thoroughbred X Canadian breeding.

After standing at stud in Maine and Prince Edward Island, Canada, he found owners in Nova Scotia in 1868, where, before his death in 1871, he covered 106 mares. He was last owned by John Hall, of Lawrencetown, Annapolis County. In 1892, a trotting son of this French bred pacer, out of a Morgan mare, was sent to Sable. He was the 14.2 h.h. chestnut stallion, Tom Thumb. Boutilier, remembering his wonderful sire, renamed him Flying Frenchman. Having spent much of his life racing, doing his mile in under three minutes, and at stud, this old fellow still showed his good mettle by more than holding his own in sparring matches with the ambitious native colts. Flying Frenchman has, incorrectly, been recorded as having "gone mad". This is one of the myths of the island and is baseless. He served well and was not put down until 1898.

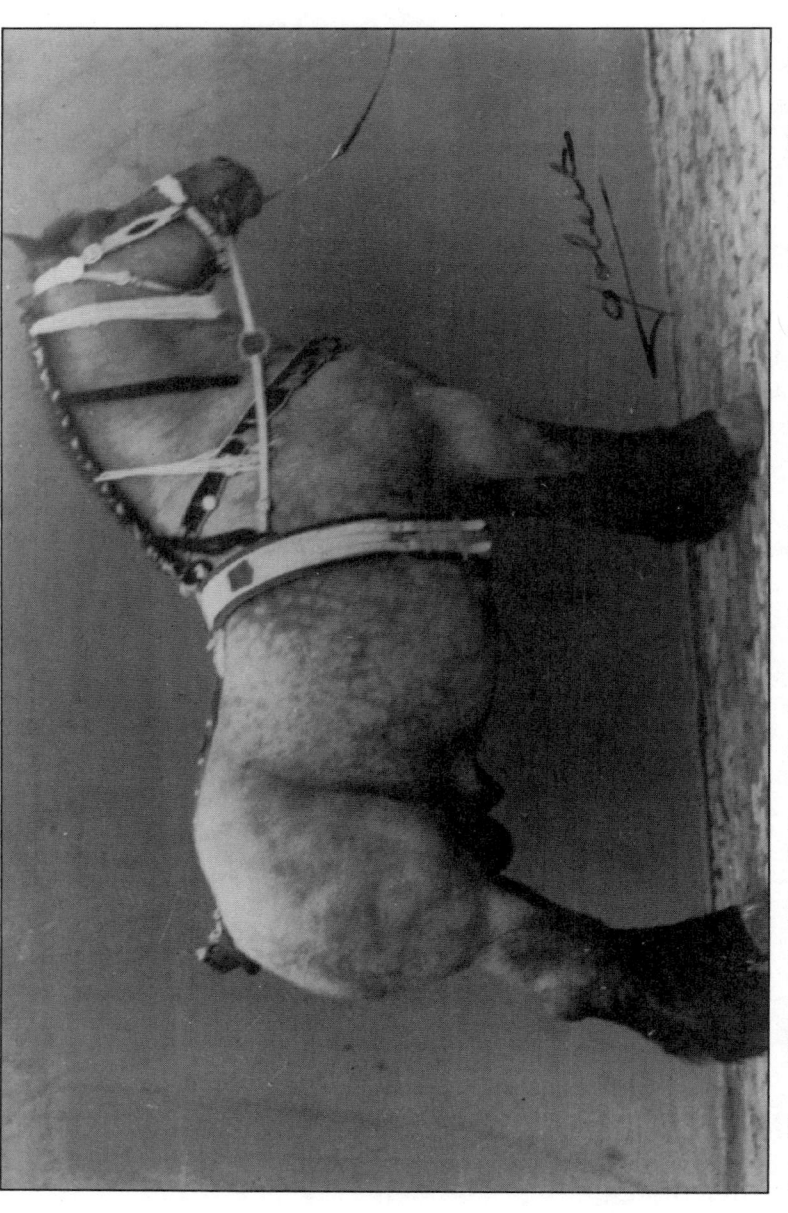

A stallion of Belgian draft breed. Bayard II, imported from Belgium, arriving on Sable Island in 1903, would have been a lighter weight type of this breed. (Photo by Gobert, courtesy of the Royaume de Belgique, Société Royale "Le Cheval de Trait Belge.")

The myth of Flying Frenchman having lost his wits may stem from a confusion in names. The story better suits the pure bred French Canadian pony, Sable Prince. This stallion was unhappy with his new home from the outset, and returned again and again to his point of landing. The Sable horses rejected the lonely colt, knocking him about, causing Boutilier untold trouble in patching him up and giving him a new location. After several false starts the homesick colt more or less settled down with a gang of his own mares, but his heart was not in island life and his approach to family duties and responsibilities were somewhat halfhearted. He is not listed after 1902, but as his purchase price had been high, he was probably returned to the mainland.

Also imported in 1900 was Pretoria. He came from Prince Edward Island (banished would be a more suitable word) and was possibly of the then popular Clydesdale X Thoroughbred type. He soon gave ample reason for his removal from the "Garden of the Gulf". With barely a few weeks to his credit on Sable Island, he attacked and demolished the Superintendent's pride and joy, a newly refurbished and varnished buckboard. All but impossible to catch, bolting if ridden, a fence breaker and general troublehorse, he was gelded in 1904 and put down the following year. When not raising the devil, Pretoria was an excellent stallion and what stock he left was of fine quality, although no notes were made as to their temperament.

During the years 1901 to 1905, an extensive program of planting and general experimentation to improve the productivity of the island was implemented by the Dominion government. Tussoc grass was imported from the Falkland Islands where, in 1834, hundreds of feral horses and cattle were reported as flourishing. Along with thousands of mainland saplings and shrubs, many pounds of grass, clover and other seeds, some eight thousand plants were brought from Brittany and Normandy. Studies had been made by staff of the Central Experimental Farm in Ottawa of similar sandy and exposed areas in Europe, and two horses of a heavier breed known to thrive in such conditions were also imported and sent to Sable in 1903.

They were of the Belgium Draft breed. The stallion, Bayard 11, was a three-year-old bay of 850 kg by Louis de Bierset (63480), out of the mare Baille (7615), etc., and was bred by L. Nicolay of Autelbas, a mountain area of Luxemburg. The mare was a three-year-old of 650 kg named Reine des Bruyers; she was by Pirate (8878), out of Niobe (9667), etc. and she was bred by Hipp Meese of Wijneghem, a sandy area near Antwerp. Both were prize-winning animals in Belgium. By 1907 nearly all imported plants were dead and Reine was to die that year along with her unborn foal by Bayard.

Both these animals were of a kind and generous nature; Bayard being miserable when taken into winter quarters. After suffering a number of injuries while attempting to return to his mares, he took command of the situation, and as next winter approached, he presented himself and family at the Main Station gates. The equally generous Boutilier took all into his care. Bayard was last listed on the island in 1908. He was sent to Halifax for veterinary care in 1909, and died there the following year.

The stallion Ottawa followed these two in 1904. All evidence points to him being the 15-year-old son of American bred, Hartford by Harold, Harold by Hambletonian. C.R. Bill of Cornwallis, Nova Scotia, had imported Hartford from the United States in 1878 for stud purposes. The last listing for Ottawa is 1907.

Basil, a yearling of unknown breed, and even less history as to his fate while on Sable Island, arrived on the Lady Laurier in 1905. He may have been a Mustang; those horses were being sold in the Ottawa area at that time. He is again listed in 1908, but when Boutilier left the island reliable records are once more difficult to come by. It is known that a large hunter named Colonel, was also a stock horse there about 1931, and Donald Johnson, one of the later "governors", recalls another stallion, Starlight, weighing some 1200 pounds, as having been sent out from the Nova Scotia Agricultural College between the war years. Neither one can be identified, nor can an estimate be made as to their precise breed.

As well as mare Reine des Bruyers, other mares were sent out to the island, including Empress, a grey Arabian, sent as a gift to the island from the City of Ottawa in 1903; Seven Hackneys, imported from either Britain or the United States, and sent to the island in 1904. As with the stallions, all mares ran freely with the native stock. If covered by a "tame" stallion, dates were recorded along with the resulting foals or losses. Such stud entries include a number of the better island mares being bred to one or the other of the imported stallions and these foals were nearly always given the name of a particular wrecked vessel, an important person or some world event. These entries, along with other information in the Boutilier journals, make these records all but priceless in the history of the island.

Chapter 7
The Old Patch Mare

The care of imported stock and that of the horses has varied a good deal: the early and latter periods leaving much to be desired in the way of good herd management. During these years stallions would often be turned loose simply "sharp shod in front", and left to take their chances. Little imagination is required to visualize the punishment they must have inflicted on the opposition. However, the island stallions are tough and dirty fighters, and were by nature more than a match for the imported mainlanders. At other times it has been said that some stallions were kept at Main Station and the island mares brought there to be bred.

But from Robert Boutilier's journals we know that after the loss of Jack Trumps, a much firmer hand took charge of all the horses, both native and imported. Selective gelding and culling resulted in a better ratio of mares to stallions and less harrying on the studs by ambitious youngsters. During his term of office all imported stallions ran freely with their mares, but these valuable animals, along with numbers of better quality Sable Island mares wintered in the large barns at Main Station, or in the smaller barns of the various stations. However sparse, the early grazing animals would frequently refuse their stable rations as soon as they sensed the growing grasses of April, a situation that forced the hands of the islanders and sent the stallions out to the herds before the mares were fit to receive them. All imported stallions were rotated between the east and west end horses, sometimes changed during the season or alternated by years. Those with a high degree of thoroughbred breeding did seem to spend the greater part of their days in the west end near Main Station, a situation which kept them more immediately under the superintendent's eye.

Although running free with their mares from spring until fall, these studs were constantly checked by patrols circling the island on the lookout for wrecks. Any wound, sickness or sign of trouble was immediately reported and quickly attended to. When old, or of little further use, they were either shipped off or put down. There is no record of a covering stallion converted to use as a draft or saddle horse after his successful tour of duty.

In winter quarters, all mares were in straight stalls or community pens, but stallions enjoyed the freedom of loose boxes; each having his name over the door or on the jack-knife gate. All were on wood standing for the early years, but later, when greater numbers wintered in and health problems arose, this was changed to concrete. Deep litter bedding was general, this being reduced in March and completely removed to the fields and gardens when the barns began to empty in April.

Care of feet, teeth and good general hygiene was practised, although, despite such attention the need for some clipping and treatment with nicotine washes reflects crowded conditions and the unwise housing of fowls in an adjoining barn. Cleansing of barns and treatment with whitewashes and disinfectants, ventilation, skylights and as much fresh air as the island's winds would allow, were all put to healthful use. The barns were banked with eel grass and window shutters were installed as protection from the wind-driven sands that will penetrate the smallest crack, ream putty from windows and leave all exposed glass etched as if frosted. A journal entry for February 1886, reads "Worst day of Winter. 10 above at night. Blowing sand and snow in clouds. Horses led to drink from barns. All under cover tonight."

However, Boutilier recognised the importance of exercise in bringing his stabled animals safely through the winter months. All stock was turned out in the yards when weather allowed and stallions had as much regular beach exercise as possible. In true island style the problem of exercising the horses singly was solved by the creation of a hitch which allowed several horses to be exercised together. December 1886, "Made a bar and fittings to work three horses abreast in truck." In 1898, and later in 1904, Boutilier carved and assembled a three-horse yoke used by teams of horses rushing the lifeboats to wreck sites.

The island grasses were early recognized as making hay of poor nutrient value. James Morris attempted to improve the quality by scratching the surface of mown fields with a home made harrow of heavy oak plank set with large wooden spikes, then broadcasting a mixture of Timothy and clover seeds.

Areas were ploughed and set for "English" or cultivated hay from then on and while earlier journal entries list loads, later harvests are measured in tons. An entry on August 6, 1900, records, "11 loads of cultivated hay, 2 loads of green oats, 7 loads of Wild hay". April 29, 1902, reads "Seed patch in Home field: Timothy, clover, oats. Seed patch at top of S.W. field with Pretoria Wheat, Riga Wheat, Mensury Barley, Early Gowa Oats, Bokhara clover, Brome grass, Timothy, clover and oats."

Main Station had over ten acres under cultivation of various kinds, other island stations produced much smaller crops, nevertheless, in 1913, the total harvest for the combined stations was 12 tons of English hay and 120 tons of wild hay. From Robert Boutilier's time until the late 1920s, an average of 40 working horses, some 60 or so head of cattle, various colts and some island mares and a few hogs had to be fed while wintering in the barns for 4-5-month period.

However, all island crops, with the exception of the bountiful native cranberries, were uncertain; storms and high seas could saturate the fields with salt water, early growth could be buried under drifting sand or sheared off at ground level as if cut by a scythe or sand-blasted. A journal entry for August, 1905, is typical: "The storm today has covered the grass with sand and broken and blackened all the vegetables and potatoes." Supplies of island hay had nearly always to be supplemented by shipments from the mainland. Wild hay was used as rough feed, at times to feed the native horses, sometimes for bedding and at others, when cut green, fed to penned "soiling cattle", animals being fattened and who produced valuable manures for the gardens and fields.

Through the pages of the various Sable Island Superintendents' reports and a host of other memoirs and documents, pictures of the island's horses come down to us in all their colours, escapades, health, adversity, changing types and their fortunes in new worlds. In journal entries such as these of February, 1887, the names, particulars and offspring of many of the native horses are found: "Run into the Pound the Old Patch mare and 2 colts." A day or so later: "Handling Old Patch mare so as to bring her into the barn." We meet the red roan mare Big Red, and the light bay mare, called by Boutilier, "The Yellow mare", who annually brought her new foal to Main Station for his approval. One mare, who was the pet of all who knew her, "Ran a mare and young foal in pound. Caught her and put her in the barn. She got very quiet and apparently tame. Named her Countess." This native mare was one of the chosen who wintered in, but her privileged way of life and the affection she enjoyed could not prevent her death during foaling two years later.

The two stallions named Chief, one sent to St. Paul's Island off the coast of Cape Breton (three in all sent there over the years) and one other of the same name, were not so fortunate. With the arrival of one imported stallion, this last was shot in order that the mainlander could have his mares. Sultan, another quality native stud wintered in and, as with the imported stallions, was sent to run for alternating periods at the east or west end of Sable Island.

Native animals were named after a shipwreck, an occasion, or someone of importance. Mira, a bright bay mare, whose son Bobs was named after Britain's soldier-hero of India and the Boer War, Lord Roberts. Coronation, another son of Mira's celebrated the crowning of Edward VII. Stella Maris, Crofton Hall, Aldebaren and Topaz, were just a few of those who carried the names of wrecked vessels. The gelding Gabrielle (named for the captain of a French vessel), Viscount Mileux, Victoria, General Middleton and Marconi all carried honoured names. Some lesser named animals were Mistake, Chum, Raff, Spot, Pat, Mike, and Midget, the beloved mount of Robert Boutilier's tiny, 4-foot 10-inch daughter, Trixie.

Besides these ran the nameless ones who appear briefly in journal entries, such as that of Darby's of December 10, 1846, "Skinned the fatt (sic) mare that we found dead (near the house) She was busted inside which caused her death." In March of 1892, "Had a crippled yearling shot in West End gang." The care taken of island stock is illustrated by the entry of April 1896: "Ran in two young horses, half starved. Put them in loose boxes and feeding them." September 1887—"Went to Big Root and caught mare with crooked and out-grown hooves. Sawed them off and trimmed them up as well as possible." The same for March 1906 — "Yearling filly found helpless at Eelpond and brought into barn." The mischief and merriment of children and horses is shown by the entry of April 1900: "The girls and boys caught a three year old black pony at Whalepost and brought him to the barn." And that of February, 1906—"Wild pony caught in wireless field wire and took a piece away fastened to its tail." The island horses will explore everything and, if possible, eat it.

Chapter 8
Attar of Roses

Horses shipped off Sable Island by the first two superintendents were simply sent as rounded up; in later years many were gelded. Some were gelded where caught; others were driven to Main, or some other station where the operation was performed. The animals were then penned until well recovered, after which they were returned to the herd. Geldings required for shipment were driven into the pound in the general roundup, where they were picked out and sent off as the market required.

After gelding, those animals to be used for island work were frequently fettered, or fitted with a long chain and trailing log called a clog, so as to be more readily identified and secured. As many of these horses were not stabled, but simply caught up when required for any particular job, they spent most of their time thus restricted. The story is told of one clever old fellow, who, when tired of his trailing wooden shadow, took long walks in the sea where he enjoyed a temporary sense of freedom as he floated his clog.

The round-ups which commonly occurred twice a year, were a busy and exhausting time for all. August 22, 1842 — "All hands after horses. We had a great deal of running today and only got five. We kept them in the pound all night." The long chases up hill and down in the deep sandy going, [type of surface travelled] often ruined or lamed the riding horses. One loss is recorded when 64 head of "wild horses" were driven and shipped on July 27, 1895. Boutilier enters in his journal, "Pony Bulgar died at No. 3 [station] from excessive running, which resulted in spasmodic colic."

Once the horses were in the pound, the dangerous task of selection began. During the later years the method used was a looped lasso on the end of a long pole. The loop was dropped over the head of a horse and the line taken up by the men, whose unfortunate job it was to bring the animal out. Milling horses and flying hooves took their toll and the superintendents were frequently called upon to make running repairs on their men. September 4, 1891—"Stephen Smallcombe got his leg badly damaged while snaring ponies in the pound."

Once the selection was made, those chosen were retained in the pound, while their more fortunate brethren were released. If

Lassoing prior to being thrown. (Courtesy, R.S. Bouteilier)

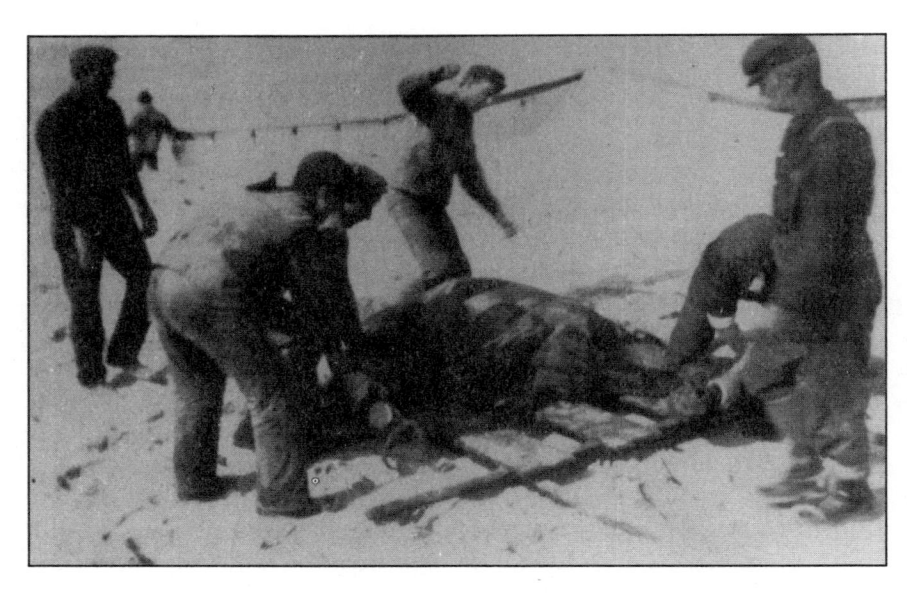

Tied to frame or pony barrow prior to loading aboard surf boat. (Courtesy, R.S. Bouteilier)

Off-loading Sable Island horses at Dartmouth, Nova Scotia, 1918. (Courtesy, Nova Scotia Museum)

they were to be shipped immediately they were roped in a very painful but effective manner so as to insure their moving forward.

When the animals neared the waiting surf boats, each was thrown and all four legs were tied together, then heaved onto a four-handled stretcher or "pony barrow" and from this slid into a surf boat. Then, with five or six others taken out to the transport vessel anchored offshore.

During the middle 1800s the horses were hoisted aboard the vessels by slings; in later years they were simply lifted aboard by their tied feet, but slings were used to bring them ashore at Halifax. Considerable care was taken when the legs were first tied to keep damage to a minimum. A soft rope padding was used and a special method of roping.

Some superintendents held the horses in the pound for a night before shipping; lacking food and water the animals became more manageable. Robert Boutilier liked his animals as strong as possible for the short voyage. He shipped his horses via the pound the same day as roundup, or fed them if weather delayed loading. He also saw to it that ample supplies of freshly cut grass or hay were awaiting them in their pens aboard ship. There, free from their bonds, "they were soon on their feet and eating."

The government snow, *Earl of Moira*, brought off the first horses for Nova Scotia in 1803, but local schooners soon began to make the run, taking out supplies and returning with stock, salvage and island produce. Schooners such as the *Wave*, the *Alice* the *Pheobe* and the *Daring*, carried the horses in deck pens, or in their hold; steamers of more modern times carried them in the hold. The *Lady Laurier* (1902-1960) was equipped for marine cable laying and had a circular tank, or hold, in which cable could be stored. Hundreds of horses were transported to their new lives in this area. Some accounts have the animals tied by the head, others, that they were untied but quite closely packed. They stood on fairly deep litter of island hay or straw.

A fast, smooth passage to Halifax or Dartmouth meant animals in good condition and higher prices, at one of the many wharves where the horses were auctioned. A rough trip could result in conditions which prompted uninhibited observations such as these in the rough journalism of 1856, from *The Acadian Recorder*, October 4.

Talk About Town
The talk is, some people are now adays reminded of the marvellous story told by the lying captain, in Peter Simple,

"Journal of a Sailor from Midshipman to ..." Capt. Frederick Marryat, 1834, author of "Mr. Midshipman Easy", of a ship on board of which a cask of attar of roses was staved in, and the fragrant liquid so permeated and impregnated with its odor every stick of timber in the ship that men could not live on board any longer and the vessel was afterwards stripped, taken to pieces, sawed up and sold for rosewood. Well, the talk is, the Board of Works has lately taken to lugging cargoes of dirty ponies up from Sable Island, in the revenue cutter *Daring*; and the effect upon the olfactories of those who have to breathe on board that vessel is nearly as powerful as that experienced in the ship wherein the attar of roses was spilled—only that the quality of the smell is different—decidedly so. The talk very generally heard about this matter is that the *Daring* cost the Province too much to be soon converted into a stinking horse boat and thus rendered unfit for any other purpose. Some people who talk upon the subject say that if the *Daring* had happened to be built by a different contractor, she would not have been sent to all sorts of places and employed in all sorts of uses as if to try to knock the craft to pieces. The talk of shipbuilders is, the sooner the experiment is concluded, the better—"all the better for trade, you know."

Politics played an important part in the everyday life of Nova Scotians in the 1800s and, depending upon the bias of the newspaper reporting the arrival of the horse shipment, the animals were either approved or damned. Whatever opinions might be, their landing and sale always drew a crowd of "experts", thrill seekers and purchasers with various sized pockets.

The Halifax Daily Reporter notes the arrival of an August shipment in 1873 with caustic comment.

Sable Island Ponies
A Sale at Conlon's stables. Prices realised were from $16 to $80. It is to be hoped that no more of these animals will be brought to Halifax. Our present breed of horse is, in all conscience, low enough without forcing these degenerate animals upon the country.

The Morning Herald of the same date cheerfully prints,

Per Order of the Department of Marine and Fisheries. Eleven ponies just arrived from Sable Island. These are decidedly the best lot ever offered. All are first class.

Thrills, some sad and sorry, are reflected in two 1879 papers: *Morning Chronicle*, September 16,

> There was quite a lively time at O'Connor's Wharf yesterday, landing the Sable Island ponies brought by the *Newfield*. The little animals were quite untamed and some of them very difficult to manage. Mr William Guinan received a nasty kick from one, which compelled him to give up work. Another man named Williamson was badly kicked about the face and chest.

Some three weeks later The *New Times Reporter* chirps happily,

> The Sable Island ponies, sold a few weeks ago, seem to have got into harness in a wonderful short time. They are docile and their owners are already making them useful.

Well, perhaps not all of them. *The Halifax Daily Reporter* covers in colourful detail the rampages of one on Barrington Street. After dumping its rider, this less tractable son of Sable charged through a plate glass window at his own reflection and after demolishing a mangle, displayed for sidewalk sale, showed a clean pair of heels as he made for quieter Richmond. "There aught to be a law," scolds the paper, "forbidding wild animals from going at large."

It is true that many of these imports found poor owners and tragic ends and that most were sold as fishermen's horses, or ended their days working in a coal mine or a plantation. It is equally true that countless numbers of mainland horses endured the same fates in those harsher days before concern for animal care and health became of interest to agricultural bodies and prior to the creation of the Society for the Prevention of Cruelty to Animals.

Barely mentioned are those who were more fortunate and who found good and caring owners. Taken for granted at the time and now forgotten are those who, smartly turned out, drew basket or tub cart, sleigh or small buggy; those stallions who left the island and, hardy and tough, gave added stamina to many of the common natives of other areas. There were also those who, on occasion, paraded with the best at shows or exhibition, and some, who to this day are remembered with awe (for they could be headstrong) or affection by those who once owned them.

The British government eventually made grants towards the upkeep of Nova Scotia's Sable Island Humane Establishment, but its cost was still considerable to the province. It was therefore important that wherever possible the venture should help pay its own way.

Cranberries grow in profusion there and hundreds of barrels of excellent fruit were shipped off to the United States and to Canadian consumers. Store cattle were sent out from Halifax, fattened, and then returned to local mainland markets. However, other than the constant shipping of salvage materials, the greatest export was that of the horses. Alive, these animals were primarily used as draft animals; dead, their hair, hides, and hooves came off the ships in barrels to be processed for upholstery and plastering needs, for leather and for making glue.

Depending upon the market and the size of the herd, shipments of horses have varied greatly in numbers and frequency. Some years show a low figure of 30, others as high as 150; some years saw no shipments at all. In years of good management as much care was taken of them as possible and efforts were made to keep the scattered groups healthy, particularly during the long, difficult winters. During the winter of 1890-1, 75 head were lost, and it was estimated that provincial coffers were thus short $1,500.

As with all wild or feral horses, in climates made up in part by long periods of cold and damp, the highest mortality rate occurs among the weak, the very old or young, among in-foal mares and the early or late born foals.

About the year 1881, Superintendent D. MacDonald built a large "poney shelter" in the hope that the animals would take advantage of the haven during storms, and so cut down on the winter losses. The structure appears to have been designed by one more used to the building of boats than the care of livestock. It was 12 feet wide and 79 feet long, with no overhead light and entry only from either end. The horses avoided it assiduously. It must have seemed to them a dark and narrow trap with barrelling drafts.

Ten years later Robert Boutilier, understanding the wild nature of his charges, built a number of smaller, three-winged shelters, all open to the sky. The three sheltering wings radiated from a central crib which was stocked with island hay during the winter months. These shelters were patronised, but not with total success, even though they were scattered about among the home ranges of numbers of gangs, or family groups. Besides the need for periodic restocking with hay, their maintenance was costly, at least in time. The horses did a good deal of damage by their rubbing, scratching and rough-housing and the timbering was buffeted by wind and silted up by the restless sand.

There may also have been a question of just how many animals were able to take advantage of what seemed to be a good idea. The pecking order seen in most animal behavior might have allowed dominant gangs or individuals to drive away lesser horses, so that the shelter and supplementary feed went to the strong, while the weaker, more needy animals still suffered winter hardship. They seem to have been in use for a number of years but after 1900, all record of them ceases. Without this help from man the horses would have returned to their old type of rough winter forage and for shelter to the lea of sand dunes until April returned the green grass and more charitable weather.

Chapter 9
A Natural Order

In severe weather the horses suffer similar stresses to those experienced by the pony breeds of Britain, but without the benefit of trees or high shrub to shelter them. These conditions make great demands upon the herd stallion who, while he will drive his family into the shelter of the high dunes, stands out in the full force of the storm, never relaxing his guarding position for a moment. During these periods none of the horses feed. Nothing is asked of the mares and foals, but the stallion must be constant in his efforts to drive off young colts who would steal a mare when the moment offers. It is a task which depletes the strength of some studs, yet the health and safety of his family depends entirely on his intelligence and stamina.

Young colts who manage to steal a mare or two, or inherit an old one, will often lose them during the winter months through lack of intelligent maturity. Failing to provide adequate shelter for them, or driving the mares too enthusiastically in spring, often results in injury and at times death, particularly to the foals. The young equine king may find himself deposed by a tougher fighter than himself, or lose his lone mare through the crafty maneuvers of another. Many battles are fought and many lessons learned before a Sable Island stud reigns supreme over his family and home range.

The herd rules that still govern the diminishing numbers of wild horses have been re-established among the feral horses of Sable Island since man has returned their fortunes to nature. The small herds, or family groups consist of a stallion, his mares and their offspring of various ages. Colts are tolerated until they show interest in the mares, then they are driven out to join groups of bachelors, known as "Rowdy gangs". Some colts show no interest in the mares and are sometimes allowed to stay within the family group, where they take pressure off the stallion by helping to drive away intruders. These neutral animals are known as "Tags".

A stallion rarely mates with his own daughter. She is usually driven out of the herd and may join other groups or be picked up by colts when she reaches maturity. This banishment is often carried out by the kicks and bites of the herd mares. Fillies are not covered with success until their third or fourth year. Foals are born at all seasons of the year with the exception of January. Most are born in May and June and during their first cold season grow thick, woolly coats. Dr. D. E. Welsh, ("The Life of Sable Island Horses", *Nature Canada*, April-June 1973) found these coats to be about three inches long. These foals, along with "mares and younger animals [are] slower to moult [and] in July the yearlings and two year olds are a motley mixture of glossy summer coat and bedraggled tufts of winter hair."

Similar behaviour patterns have been noted in studies made of some of the world's free ranging or feral horses, such as ritualised grooming, the ritual attention to manure piles when two groups meet, or when one stallion visits or challenges another. The full display pattern of greeting, acknowledgement of superiority and challenge, as seen in the Sable Island horses, is not commonly witnessed in the British breeds, as several, such as Exmoor, Dartmoor and Welsh, are still subject to annual drives, or drifts, which remove for sale most of the entire colts.

Stallions among the British breeds run with many more mares than do those on Sable Island, who run with from 3 to 20 mares and their young. The total number on the island (about 250) is divided between 40-50 stallions. The full family structure is incomplete in the British breeds. The social position and role of each member of the family group is firmly established and adhered to, particularly with regard to the stallion and senior, or dominant herd mare. While it is this mare who leads and decides when and where the group feeds and waters, it is the stallion, or Sultan horse, who drives the group from the rear when he feels them endangered. Any attempts to alter the social role by a member, or to usurp the position of the stallion or herd mare may result in the animal being driven from the group. Where the aspirant is successful, the mare may take a secondary role within the herd or be taken up by the ever ambitious colts. The deposed stallion then leads a lonely and solitary life, often growing an excessively long mane and tail.

Chapter 10
Golden Horses of the Queen

In addition to the first superintendent's description of the horses in 1802, an assessment of them appears in the lecture given to Nova Scotia's Institute of Natural Science in 1864, by Dr. J. B. Gilpin.

> The head large and ill-set on, with usually the round Roman nose and thick jowl (see drawing of charging stallion) the ear small, short and square at the top; crest very thick and heavy in the male; neck cock-thrappled, or swelling out in front; withers very low; quarters short and sloping; legs very strong and robust, with thick, upright pasterns; the eye not large or bright; the mouth very short; the forelock and mane abundant; the mane nearly to the ground, the forelock covering the nostrils; the weight of the mane often pulls the crest over, so that in the mares, the neck seems ewe-necked; the tail also nearly reaching the ground; the fore toe usually turned outward, or paddlefooted; the withers seemingly lower than the rump or quarters ... the coat is, during winter, long and shaggy, especially under the chin and on the legs.

A modern description of them is given by Dr. D.A. Welsh, Department of Biology at Dalhousie University in "The Life of Sable Island's Wild Horses", *Nature Canada*, April-June 1973.

> A typical Sable Island horse stands about 14 h.h. (4 feet, 8 inches) at the withers and weighs from 500 to 700 pounds. In appearance, they most closely resemble the Barb of North Africa. They tend to be chunky horses; narrow, deep chests, heavy shoulders and necks, and short, heavy legs are typical. All Sable Island horses have small, rounded wide-set ears that are slightly tipped inward, and many have fine muzzles and thin, curved nostrils. Roman noses are common, although a few have finely-dished faces. Their hooves are small and round, but often become overgrown in the soft sand. The stallions have exceptionally long forelocks, manes and tails and in extreme cases mane and tail may even reach the ground. Manes and forelocks of mares are relatively short.

Stallion and mare showing Roman noses. (Courtesy, Dr. D.E. Welsh)

Mare showing dished face and thick lower leg. (Courtesy, Dr. D.E. Welsh)

Four observers give the colours of the horses. Morris gives two descriptions the first: "They were in general bay coloured, a few black, only two iron greys and two pied white and dark bay [skewbald]"; the second, "I expect three quarters of them are bay-dark and bright and the others various coloured—some reddish-black pied, and whitish and grey—very few of the latter." Gilpin, said in 1864,

> I found the bays to be the most numerous, including the brown with them, and the rest chestnuts. Of black there were few. Of greys, none. A peculiarly wild mare seemed to be a red roan ... there was one pure white young horse that must have been foaled white from his age. Of piebald, they had so run into the colour that means had to be taken to lessen them by destroying them and by sending them off the island; and lastly the bluish mouse colour, often when a black stripe along the back, seemed to be nearly as numerous as the chestnuts. Many of the chestnuts seemed running into Isabella [Isabella (Y'sabella): Bred since the fifteenth century when Queen Isabella of Spain formed a stud of coloured horses for herself and her ladies to ride; some of these 'Golden Horses of the Queen' were given by the Queen to America, which was then being colonised by the Spaniards. Originally from Arab stock, they vary in colour from light to liver chestnuts with white mane and tail, more usually known now as palomino [Summerhays' Encyclopedia for Horsemen] but I saw none with black lines around the legs, and but few with black list along the spine.

Dr. Welsh states in his "Population, Behavioral, and Grazing Ecology of the Horses of Sable Island, Nova Scotia" (1975) (when the average size of the island herd amounted to 250 head),

> Almost half of the population are bay, 22.5% are chestnut, 13% are palomino, 9.5% are black, and 8.4% are brown. They have a full range of white face and leg markings that are common on brown, black and palomino horses. Many have a dark dorsal stripe and some have shoulder stripes. Almost all have a light-coloured "mealy muzzel."

In 1864 Gilpin wrote that the piebalds were being sent off the island, but they were to remain for some years more. Twenty-one years later the Rev. William Almon DesBrisay, "resident missionary on Sable Island", gave the colours of the horses as "light chestnut with a mane almost white; jet black, red, mouse color

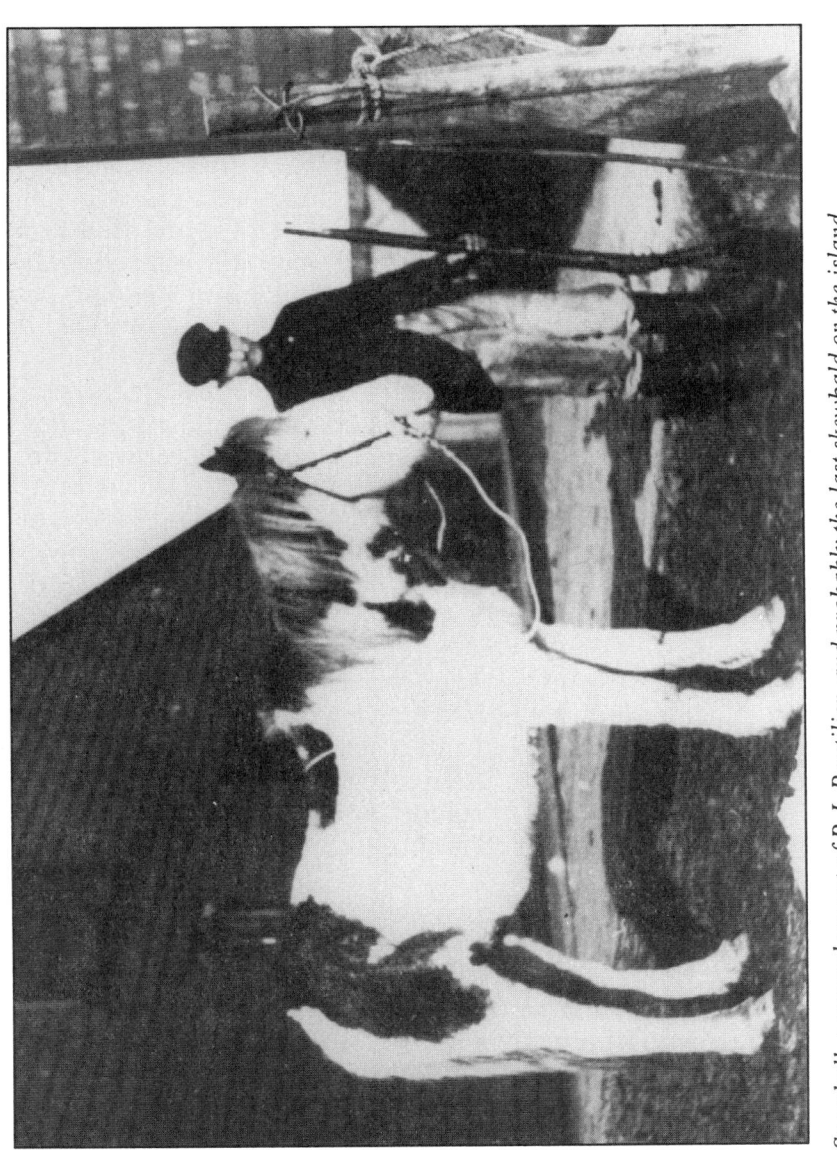

Snowball, personal mount of R.J. Boutilier, and probably the last skewbald on the island. (Courtesy, Miss Ruth Tobin)

spotted, brown, and bay and one piebald or ... patchwork horse." DesBrisay also notes the long mane and tail, and terms them: long enough for blankets and nightcap." He also records, along with Gilpin and Welsh that the " 'wall eye', also called Blue, or China eye, is occasionally to be met with."

In Robert Boutilier's journals these colours are to be found; bright bay, dark and yellow; chestnuts, and those now incorrectly termed Palomino; red roan, mouse, brown. Other writers and sales advertisements refer to a colour which is given as "battleship grey", grey, and blue roan, but this is possibly the same colour seen which most call "mouse". Boutilier, riding 14 stone, seems to have used the last piebald, or skewbald horse for his personal mount. Predominantly white, this unusual gelding was named Snowball, and, along with the imported stallions, enjoyed that name over his stall in the barn.

A number of colours are no longer found among the horses; they have been lost due either to a lack of popularity, or the reverse. Yellow bays, (no yellow duns or buckskins have ever been recorded) the greys, mouse, (probably the same listed as blue duns) through lack of markets, The yellow bay was in disfavour as his colour, or lack of depth to his colour, was thought to denote weakness in a horse. The greys and mouse also found the poorest purchasers as their colour and coats were the bane of good grooms and owners. It is difficult to keep a grey looking clean and it is difficult to get a shine or bloom on this colour horse. The skewbalds and piebalds found a limited market for butcher's and other tradesmen's carts, or for children's mounts, but found no favour for use in coal mines due to their carrying so much white. The island herd began to "run to colour" and with few markets for them, they were either sent off and sold cheaply, or destroyed on Sable Island.

Of their gaits, we have few descriptions. Morris notes their "having in general a handsome trot and canter." This fairly high action may have come naturally to them, or it may have its origin in the deep sand of their range which makes them lift their feet today. DesBrisay, as given before, wrote of their ability to trot, gallop, paddle, rack, prance, shuffle, waltz and pace. While the reverend gentleman seems somewhat carried away by the varied actions of the horses, his description does give an impression of animals whose blood stemmed from ancestors capable of High School airs.

There are several occasions in past years when visitors to the island have noted the ability of the horses to pace; this has also been said of the Acadian French horses, yet many of both breeds appear to have been double gaited. Cape Breton once produced a small, fast, round-gaited trotter and Frank Forester, in *The Horse of America* (1857), wrote that the Canadian horse had a "fine, high action, bending his knee roundly and setting his foot squarely on the ground". Yet it is a well known fact that many of the horses of Quebec also paced long before the American Narragansett pacer began to arrive in there. In all probability trotters and pacers were to be found among all the French horses in New France, depending on conditions, needs and preference. Some visitors to Nova Scotia wrote that all common horses there paced. However, some of the early roads were so rough as to make life difficult for a pacer to move at speed. Others remarked that the horses were famous for their high trotting action, caused, it was said, by the youngsters spending their days grazing among rough stumps and rocks which made them lift their feet. If the Sable Island horse has connections with these mainland equines, it is not difficult to believe that trotting, pacing and other actions could have been seen among them. The author has had little opportunity to observe their action for herself, and those who have spent some time on the island in recent years have made no note of any in their studies.

Chapter 11
"Inured to the Spring of an English Hunter"

People living on Sable and those visiting, have often given their impressions of island life; not a few have left us their opinion as to the origin of the horses. The list is interesting.

They are said to have been taken to their sandy home by the Norsemen, or Cabot, by Baron de Lery in 1518, by Portuguese fishermen in 1582, by the Marquis de La Roche-Mesgouez in 1598, and by Acadian horse ranchers in the early 1700s. They are said to have descended from those escaping from an undated, unnamed Spanish wreck, from those surviving the wreck of a vessel out of Le Havre, France, that was on its way to Quebec, again, undated and unnamed. They are said to have their roots in horses being transhipped from Massachusetts in yet another unnamed vessel, said to have been wrecked on Sable Island around 1800. The history of the island and the possibility of some of these fables being true has already been given. The readers must judge for themselves the probabilities for the remainder.

They have been likened to an equally wide range of the world's horses. Observers have seen in them the North African Barb, the Siberian Wild horse, the Tarpan, the Spanish Barb, Mexican pony, French Canadian pony, the Scottish Garron, the Orkney pony. One observer in 1864, when looking at the "large head, thick crest, cock-thrappled neck ('like a game cock when he crows,' says Xenophon), abundant tail and mane and low stature," saw in them the horses of the Elgin marbles. Observers have cast a wide net and Boutilier even includes the Shetland pony when he calls the smaller horses "shelties".

After a three-year study of the island and its horses, Dr. Daniel Welsh concluded the horses "are more similar to each other than to any known breed, and the unique characteristics of these horses as a whole are undoubtedly the result of natural selection due to exposure to the rigors of the environment over many generations without human interference." Dr. Welsh was looking at modern horses for comparison, as have most modern observers. A few similarities between the island horses and those of early Canada and Acadia may support the theory that the origin of the Sable Island horse lies in animals who once lived much closer to the island.

Referring to the first description of the Acadian horses as given by Cadillac in the 1600s, when he wrote of their "fine build, good shoulder, clean, strong legs (i.e., little hair) the head a little large" and the added observation that "no care is taken in raising them." Compare this with more modern descriptions of the island horses. In June of 1959, Dr. F. J. Leslie, Associate Chief, Canadian Department of Agriculture, Production and Market Branch, Livestock Division, had this to say of them, "They appear to be from photographs, to be low-set, well muscled animals, and although a little droopy in the croup, possess more substance and bone than one would naturally expect." Later that year, Dr. Leslie visited Sable Island and in September made this report,

> The ponies are of fairly good conformation but decidedly lacking in quality and are characterised by a fairly large, plain head. The majority have good backs and middles but the croup droops sharply to the tail head. All the ponies appear to possess good feet and legs.

Surely Cadillac and Leslie were describing the same type of horse?

The colours of the island horses have already been given. Now let us consider what is known of those of the Acadian, the Canadian, and those of the European Breton horse. Once more the Acadian horse evades us and we know little as to the colours favoured by the early French settlers in Acadia, but 'Gledecoloured' (sandy-grey), black and fox-coloured is referred to. Colours of horses of the same Norman-Breton breeding in Montreal and Quebec for the early 1700s run to red, black, mouse, brown and grey.

Spotted stallions are said to have been included in the early stallions Quebec received from France, and roans, skewbalds, chestnuts and greys were seen among the "countryfied, near-the-ground, well-limbed" Breton horse—the Bidet, by Earl Cathcart, in his report in the "Journal" of the Royal Agricultural Society of England of October 1889. These last were the "rustic" horses of Brittany, probably little improved since the days when these and the Norman horses were brought out to Canada and Acadia.

Arthur Young, Britain's agricultural revolutionary, in his "Travels in France", 1787-9, wrote of the "bas-Breton noble, with his large sword and his miserable but nimble nag." The combination has some echo in "Letters from Nova Scotia", by Captain Moorsom, 52nd Regiment, written in 1830. Although "inured to the spring of an English hunter", the Captain nevertheless

Kris Smith on her Sable Island gelding, Flash. "A rough ride at the trot." (Courtesy, Kris Smith)

preferred the small, tough French (Acadian) horse for his travels. Admitting his wanderings "brought their own punishment", for the roads were extremely rough (some composed of stones as large as a man's head), and "there being on either side deep bog into which a horse, if once plunged would no doubt run no small risk of remaining till domesday." However, by keeping to the middle of the track, he was "assured, by way of consolation, my horse would not sink above his knees." He much admired the way "horses bred in this country" skipped with ease along the country tracks. The same dexterity and handiness has been illustrated in the Sable Island horses. Another example of this ability, so akin to the Acadian horses crops up in 1879, when a "Superior consignment of Sable Island horses arrived in Halifax, and were advertised: 'The attention of the Officers and Horse fanciers is directed to this sale as some of the ponies are suitable for Polo.'" Polo was regularly played in Halifax between the years 1878-1905. The horses commonly used were small Thoroughbreds or part-breds, often pedigreed, who stood between 13 and 14.2 h.h., but someone must have seen in the cut, the courage and stamina of the island horses, a useful animal for the game.

He was known to be an extremely flexible animal, but with his short, upright pastern and not infrequent straight shoulder, his stride could be far from elastic. Captain Moorsom noted this in his French mounts; the shaking up he received in his travels resulting in a well-stimulated liver.

One benefit stemming from this healthful state of affairs was an ample flow of gastric juices; digestive aids which allowed him to better cope with the "sinewy hens and hard bacon" served with depressing regularity in early Nova Scotia inns.

One other reference to the French horses is contained in the account of travels made in Nova Scotia by two Yorkshiremen. In 1774, John Robinson and Thomas Rispin made these observations of the native horses. "The horses are small, chiefly of the French breed, about fourteen and a half high, plain made, but good in nature. They seldom draw with any so that few keep more than one or two for riding; they all naturally pace and will travel a long way in a day."

This last ability was noted in the horses of Quebec and in Sable Island horses in accounts given by others. DesBrisay found the island horses gave the same rough ride; "a splendid cure for indigestion." He also wrote of the horses which were part-broken, often gelded, and then returned to the herd, either fettered or free. These animals would often be rounded up and pressed into service during roundups. He found them all but tireless. Ridden over the long, punishing miles, they would oblige "at a moment's notice" with any gait and at day's end would show but little fatigue. Cathcart noted this ability when he wrote of the rustic Breton horse:

> He thrives on scanty food. *"Le cheval qui mange de tout va partout."* (The horse that feeds goes where he needs.) According to his driver, the little Bidet is "indefatigable." He runs 30 or 40 miles for days and comes in nearly as cheerful as he started.

The horses of Quebec were famous for this same quality; not only could they travel long distances with ease, but they were able to cover roads deep with snow drifts; conditions that had ruined fine blood horses.

The Acadian horse and that of Canada were basically of Norman-Breton stock; the Sable Island horse, with so many characteristics in common with the first two, gives evidence of stemming from the same root stock. Of the hardiness of all three there can be no doubt—the Sable Island horse by his very existence. Of the Canadian in early days, Vernon C. Fowke writes,

> The French-Canadian custom regulating the freedom of stock, *l'abandon des animaux* (or simply *l'abandon*), permitted animals to go at large from fall to spring, supposedly after the old crop had been garnered and before the new one could be damaged by grazing animals.

George Barnard of Sherbrooke, Quebec, writing in the May 1846 issue of the *Spirit of the Times*, extols the virtues of the Canadian horses who, rough and rustic as was their care, could, when bred to a common 15 h.h. mare, though the stud be only 14 h.h., produce in their off-spring an animal standing 15.2 h.h. He notes,

> These are the horses that yield both sport and profit to the inhabitants of Canada; and small and rude as they show, these are the horses whose blood crossed on the larger female stock of our immediate Southern neighbours has helped to give celebrity to the horses of Vermont, and produced some of the best animals either for work or speed.

Nova Scotians knew these qualities when they sent the stallions Napoleon and the Canadian pony Sable Prince to the island.

The stamina of the predominately French breed of common horses in Nova Scotia is illustrated in an agricultural report found in the 1912 Journals of Nova Scotia House of Assembly.

The author, M. Cumming, Secretary for Agriculture, writes,

> Not long ago the writer asked a farmer in this County, "What horses do you use, Clydesdales?"
> "No," he promptly replied, "I had a good Clydesdale team, but they would tire in a hard day's work on ploughed fields, and so I went back to the old breed."

The coarse feed on which all three breeds thrived is shown in the rough grazing and exposed living of the Canadian horses who were abandoned. Writing in *Colonial Farmer* of March, 1843, Titus Smith observed,

> [The Acadian horses are] necessarily small, for many of them were never fed till they were rising four years old; they procure a living in winter by gnawing upon the grass-land when they could come to the grounds and by picking the hay and straw from heaps of manure when the snow was deep, sheltering themselves in cold storms by huddling together in thickets of Firs ... Having been bred in the poorest parts of the province they make excellent horses for steep hills, bad roads and scanty pasture.

The Sable Island horses also took a delight in charging up the sand hills when ridden in round-ups. Federal correspondence concerning the island horses contains a passage by C. G. Williams, Government Superintendent of Lights, regarding an extremely poor winter when the horses would not eat the hay sent in for them "but would eat the straw bedding that was thrown out with the manure from the barn ponies." Although these horses would not eat hay which contained the scent of man, they ran true to type, for these were the horses with blood in common with those of Acadia and Quebec, who could "pick a living from anything, or nothing at all."

The various sizes and types among the early Acadian horses are not known, but those bred in the Beaubassin area (now Cumberland Basin) were said to be superior. A small type of French horse and pony were to be found in Prince Edward Island, New Brunswick, Cape Breton Island and the islands off Newfoundland. The old, fast, small, round-gaited trotter of Cape Breton was a direct descendant of the Acadian horses. All of these animals were influenced by outside blood over a number of years and that influence came mainly from the English colonies.

The size and type of Sable Island horse varies and there is no fixed type. Some animals with better quarters seem to have been there until the late 1800s. Illustrations show a horse with higher set of tail and a head much like that of the Andalusian. The paddle foot and short sloping quarter seen in varying degrees in all the horses seem to be two constants in their conformation. It is interesting to note that travellers in New England and Nova Scotia during the 1800s often wrote of the same type of quarter in the mainland horses. The paddle foot can be seen in photos of Nova Scotian horses in the late 1800s.

The horses of Quebec also varied greatly in later years, although basically of the same mixed blood as those of Acadia. Light, fast horses came from one district; heavier, slower horses from another. Some of the speediest Canadian trotters and pacers were very small— "little rats" who looked "nothing at all" out of harness, but "once put to his task" became a spectacular racing machine.

George Barnard gives life to one of these little horses in his description of ice racing in Quebec in 1842 published in *Spirit of the Times* (February 15).

> But the horses are going up to take the start. Each is harnessed rather clumsily to a low boxed sleigh, which shows him up to advantage, and gives the driver a position standing up close behind his nag, where the short reins afford him great command, and the short whip may do its office. There are five horses: here is a noble bay—what a neck he has— how round his body—how well he is balanced at both ends. He is a mongrel, which adds to his size and appearance [probably the Thoroughbred crossed on the Canadian, and known as the St. Lawrence breed of horse]. The next is a black rather heavy in the head—what large cords he has at the gambrels! [The prominent bend in the horse's hind leg] Here comes the grey—his head is light and fine enough, and his legs show blood— shaggy fetlocks, however. This is the Canadian; you may read evidence of his pedigree in that proud, cresty neck, and the terrible muscular frame, the bloodlike extremities and roomy throat are a turn up from the Arab blood of his ancestors. Now comes a mare—she walks well, is clean built, with a fine muzzle, and in good trotting form. 'But what is this drawing the next sleigh—her foal?' No,— 'Well, it cannot be more than a stunted yearling!'

That is a little pacer, one of the old Norman palfrey breed, whose race, for want of crossing, has dwindled to this size and appearance. He is of a bright chestnut (the favorite color of Jean Baptist), and has three white legs and a bald face [white with pink nose] (all the better for him). Well, now, what a wretched little, weak-built, lop-eared, ragged-hipped, slope-rumped, cat-hammed, curly-tailed creature it is! There he goes to start with those powerful horses for the mile.

And now we see them turning yonder. Presently they will be off. Now they come. Each takes a track for himself; the ice being hard and glare for a great breadth, and the brave horses well accustomed, and not afraid of the black spots in it. See, the bay and black are left behind, and the grey leads tossing his mane above that lofty crest, and springing forward, as upon legs of steel. He is a gallant horse—I think I'll buy him. But what—the mare is gaining; she moves as unerringly as a machine. And see the little pacer—his legs going like a spider's—only one side at a time. Oh, Jehu! how they come! Our favourite grey is far behind—the black and bay still farther—the pacer has come up with the mare, and every stride of the imp swings freer and faster, as he finds he is gaining on the others. The short whip is put to the mare, and she would as soon think of flying, as breaking from her trot; her croup dodges faster, and her neck is outstretched. They have passed us with a whirr—we only saw the pacer's nostrils, red, and distended, like the mouth of an India warrior giving the war-whoop, as he slide by the goal.

Such a horse was Flying Frenchman, the sire of the stallion of the name sent to Sable Island in 1892. He had the same unimpressive looks yet, for his day, he exhibited the same blazing speed and energy. No bit could hold him in a race.

The drawings of the charging stallion, and that of the horse Jolly stuck in the quicksands, contained in James Morris's report, sent to Britain's Colonial Office, must be looked at carefully; both writing and drawings have been copied by a tidier hand. As financial support for the humane settlement was always being sought from Britain as good a case for the venture and for the worth of animals found on the island was no doubt one reason for a good deal more all round refinement in the copy than would have been found in the original. Thus in assessing the conformation of the two horses, the copying must be considered.

Although Jolly was mainland bred, he exhibits some characteristics in common with the Sable Island horses. The Acadians began returning to Nova Scotia after 1762, and many would have looked for the old type of horse they had relied upon before their expulsion; Jolly was probably of this Acadian breed. Although he shows the dished face, he has the same strong build and the "cockthrappled" neck of the Sable Island horses. His ears are short and pricked, the head—again in common with the island horses—ill set on (with little throat room). His mane, although hogged (cut short) and tail give evidence of full and heavy growth. His tail is docked; a common practice in that period for many horses, particularly saddle animals. The French also docked the tails of the horses which they frequently drove in a tandem hitch, that is, one horse in front of the other. The short tail was to prevent the tail of the leader from damaging the eyes of the wheeler.

In an attempt to give the impression of quality, the tail and mane of the charging stallion has been drawn far too fine for it to be an accurate account of a horse in the rough—particularly a Sable Island horse. The head shows a Roman nose and a shape that is commonly found in the modern horse. However, both concave and convex heads are to be found in Sable Island horses, just as they were to be seen in the old French Canadian horse of years gone by.

The ears in the drawing are small and pricked, as are those of Jolly and the modern horses. The neck is extended as the horse charges but still shows the swelling front and poor junction of head and neck. The withers are low, the quarters are short and the tail typically low-set. He shows some hair at his heels, but the legs are drawn far too fine for them to represent the thick, strong lower leg of the Sable Island horse.

The story behind the incident is interesting and gives some idea of the courage and stamina of the horses; qualities they may well inherit in their Barb or Spanish blood. Morris writes:

> I have ... attempted to represent the horse that was shot by one of the ship's people [from a recent wreck] as he was hunting in company with one of his ship mates after duck near one of the ponds—about two miles from the halfway house they discovered five wild horses at a small distance one of them a fine Stallion came towards them full speed in terrific manner (which I had several times observed) one of the men made his escape behind a sand hill, the other stood his ground having a gun that seldom ever misfired, and being prepared and not dismayed at his awful and majestic appearance was determined on the fatal attempt—and not to

fire at random as you will be pleased to observe the heroic touch—the horse on receiving the breast high up shot did not die immediately although the shot brought him down, recovered and ran about half a mile, then the heart was compelled to stop its motion and by the rash judgement of the animal that possessed it. This I painted from Ideas which I obtained from the person's information who had but one eye which occasioned him to hold the piece in such a manner.

Horses in Acadia and Canada were widely used as saddle horses and one illustration of a stallion sent from France to Canada appears much like the Spanish Genette. Compare the above account with that given of the Genette in 1664, in *Le Parfait Marechal*.

> The Genettes have a wonderful active walk, a high trot, an admirable canter and an exceptionally fast racing gallop. [The early Canadians were frequently in trouble with their intendants for spending too much time and effort in racing their horses.] In general they are not very big but there are nowhere better horses. I have heard extraordinary tales of their courage. There have been horses so badly wounded that their guts were hanging out and they lost all their blood, but with the same courage and pride with which they carried their rider into battle, so they carried him out of the foray and saved his life, then they lay down to die since the spark of life had left them sooner than their courage.

The Barbs were not noted for their speed or courage; but they could endure. The Spanish horse had speed and courage to spare. Combined in the Genette, the result was formidable.

Although the expulsion of the Acadians had been pressed for by England's North American colonies, its accomplishment became a point of controversy which continues to this day. Had Hancock's animals been of the breeds left behind by the departing French, it is unlikely that much note would have been made of their shipment to Sable Island, and equally possible that England's officials would not wish to refer to those troubled days by designating the animals as Acadian in reports of later years.

Barb and Spanish characteristics in four groups of horses—early Acadian, Canadian, the horses in early North American colonies and those of Sable Island—have been considered along with the history of their area. The conclusion reached is that the Sable Island horses stem not from a Spanish wreck, but from mainland horses. They descend from animals that were interconnected with those from Canada and the English colonies that were extant on the nearest landfall which was at the time, Acadia, now called Nova Scotia.

Chapter 12
Significant dates in the search for the possible origin of the Sable Island horses.

    There appears to be insufficient evidence of support for the purported expedition of Baron de Lery in 1539.

1553   Horned cattle and pigs taken to Sable Island by Portuguese.

1598   Some sheep may have been landed by the Marquis de la Roach, Animals left by Portuguese still on Sable Island.

1633   Acadians and New Englanders begin removing and slaughtering the Portuguese animals.

1671   Nicholas Deny survey reports no animals left on the island.

1738   The Rev. Le Mercier reported that there were no animals larger than foxes actually on the island in 1737 when he and his associates put horned cattle, sheep, hogs and horses there.

1753   Le Mercier venture terminated.

1755-1762 Expulsion of Acadians, their animals abandoned.

1760   Some time before this date Thomas Hancock sent horses, cows, goats, sheep and hogs to the Island.

1760   No wrecks, Spanish or otherwise, are listed from this date from whom the modern horses could descend.

    American War of Independence (1775-1783). Both sides in the conflict raided the island and all beef cattle were removed. Both took off horses for remounts. The modern Sable Island horse descends from those that survived the raids.

    A later winter of severe cold killed off all the island's hogs.

1800   Joseph Scrambler of Dartmouth, Nova Scotia, sailed to the island on His Majesty's Tender *Trespassey* to survey and report to Sir John Wentworth prior to the creation of the humane settlement. He informed the governor that Thomas Hancock 'fitted out a schooner', and sent animals to the island, a statement suggesting local or provincial knowledge of the methods used, and preparations made, for the shipping of the animals. Had the shipment been made from Boston, or some other New England port, the simple statement that the animals had been sent is more likely.

Chapter 13
I Beg to Report

To illustrate the world in which the island horses served for so many years, the following sections are included.

In 1809, Edward Hodgson succeeded James Morris as superintendent of Sable Island, having begun his career in 1789 as seaman and serving under DesBarres while he surveyed the coasts of North America. Rising over the years to Chief Mate he served with Governor Wentworth, who placed him in his first position on Sable Island under the first superintendent.

In 1830, he petitioned Nova Scotia's House of Assembly for financial assistance:

> My salary has been so low that I have not been able to Make any provisions for a State of Inability, my whole life has been a Round of Devotion to Government Interests, having Served it now about 40 years. I am now about 66 years of Age with my Constitution very much injured by excess of fatigue, in the Water and on the Beaches, in Cold, Wet and Fog, by Night and by Day, with one of my Legs fractured and the other giving way from fatigue that I find the Service of the Establishment rather heavy upon me, in my Debilitated State, and I have to beg you Honble [sic] House, to take this my length of Service into Consideration, and be pleased to allow me to retire, granting me such Competence as will enable me to Drag through a few years of Miserable Existance, for which I will ever bound to pray.

The petition sounds, at first thought, rather as though Hodgson was coming a "bit of the old soldier" (and he carefully omits mention of the French frigate *L'Africaine*, lost off Sable Island, 1822. Two hundred French military personal were rescued from this vessel and so grateful was their monarch, Louis XVIII, that he acknowledged the brave services of the islanders by sending them a silver cup filled with gold coin and a medal struck to mark the occasion). However, any student of Sable Island history will see it rather as an understatement in an age when thousands of the poorer classes served their governments, particularly during wars, only to find themselves penniless and with ruined health after years of courageous service. Hodgson was granted a pension, but he died the year in which he applied.

Some idea of shipwrecks and the stress suffered by both humans and horses which played such a large part in island work is illustrated in the petition. Hodgson writes:

> I four times crossed the Atlantic [Sable Island to Halifax] in an open Boat, with Cast away People, exposed to every Hardship and fatigue to relieve the Island and the Establishment from encumbrance and expense, having always to bear my own expenses whilst away from the Island, and was never remunerated by government ... [I assisted at wrecks] Saving Property, offtimes with risk of my life, and limbs, often robbing my family of their little Supercargo [spare clothes] to bestow it on naked Castaways...

He then lists some of the wrecks:

> the first serious trial I had after the commencement of my superintendency ...was the Case of the Adament, Capt. Ridley, which I found in the Breakers in a Gale of Wind—waterlogged, and totally wrecked. No boat could approach the Wreck, and there were none to leave her, I did various attempts to get on board, succeeded, by Swimming through the breakers (where I was nearly drowned) where I found on her deck five Dead men and four more at the point of Death, exhausted from hunger and fatigue, those I sent on Shore, by means of Ropes, with the assistance of my people on shore, and succeeded in recovering them to life. Naked, in the Early part of the Winter, I had to Clothe them all, with Jackets and Shirts—for which they were not able to Pay and I had no other recourse but to lose them.
>
> The Barbadoes Frigate (and two other vessels in her Convoy) Loaded with money, where I was primarally instrumental in lifting all the Money out of the Water from where the Ship was Wrecked, and landing it safe on the Island, without the Loss of a Dollar, with the boats of the Establishment. In the Case of the Brig Hope stranded on the NE Bar about 1823 with 200 Souls on Board, that came on shore many of them quite Naked, to cover which I exhausted every article of Wearing Apparel that I had in my family; and at last had to Cut up the Curtains of my bed, to Cover the Poor Naked children—out of this Vessel I saved a great deal of Property, which I shipped to Halifax, which Salvage must have benefitted the Establishment very much—the Next

Case is the Ship Elizabeth, wrecked on the South Side, in the Month of January 1825, where I was exposed the whole long Winters Night and Day on the Beach in a Dreadful Snow Storm where I succeeded in Saving the whole of the Crew, and the Materials of the Ship, but with the loss of my own Health, and the breaking up of my Constitution, from the excess of fatigue, Cold and Suffering that I experienced that Night, I have never got the better of. It is my opinion, that without the great exertions used by myself and my people the whole of this Crew would have perished as they were all in a State of Intoxication.

The Bark Echo, wrecked on the South Side in the Fall of 1827 in a Gale of Wind, which lost all her boats and I succeeded with great Difficulty to get a boat off to her after upsetting several times, this Crew would also have been lost totally, but for the assistance of our boat and People we succeeded in Saving the Crew and passengers and the Materials of the Ship.

The Ship Melrose, wrecked on the NW Bar in the year 1828, where we succeeded in saving the Ship's Materials, and part of her Cargo Consisting of heavy Bales of Cotten Wooll, in the Shipping of Which, I met with an accident on the Beach that nearly terminated my existance for with a heavy Sea boat and a Large Bale of Cotten, which broke my knee Bone, and other ways injured my leg so as to lay me up a Cripple for, about 13 months and that, I am yet, in a very Debilitated State from the Injury Done my leg, and I almost despair of ever again having perfect use of it.

... the French Frigate that was wrecked on the South Side in 1822, where with the Dint of exertions by myself and People we succeeded in Saving about 70 persons that were actually upset out of their Boats in the Breakers, and would in all probability have Perished but for the kindly assistance given them on the Beach.

During the years of his service on the island the Establishment gave help to survivors, and obtained salvage from some 35 shipwrecks,

and there being saved ... to the number of about 1050 souls, including a great Deal of Property Saved from the Vessels wrecked, which has all been carefully preserved and sent to Halifax, on account of Government. We have taken every year upon average about 100 seals, producing upon average

2-1/2 tons of good Oil, which with their skins has all been carefully Preserved, and shipped to Halifax.

Boats, ropes and all lifesaving equipment as well as the islanders themselves, had all to be rushed to the site of the wreck over the difficult terrain. Pushed to the limit, the horses that carried all, had then to stand in all weather until needed for return journeys. Without their help, aid for wreck survivors would have been all but impossible.

* * *

A survey made of the island in 1848 by William T. Townsend at the government's request, gives some idea of the establishment there and the various stations existing at that time. Later years saw added stations and lighthouses at both the east and the west ends.

Schedule of Fix'd Property at Sable Island Principle Station or Head Quarters [later called Main Station.]

... situated about Six Miles from the West and on the North Side of the Island, the large Supply of Fresh Water in the immediate vicinity was the principle reason for fixing upon this Spot ... three large dwelling houses in good order, one occupied by the Superintendent, Mr. Joseph Darby, and family—consisting of Seven persons in all. The other two houses are unoccupied and are kept for casualties, the two are capable of Accomodating from One hundred to One hundred and fifty persons, and more if necessity require it. There is a large Barn and Stables combined with two large Sheds ... The whole capable of accommodating about Sixty head of Cattle and containing about Thirty tons of hay.

A Warehouse fifty four feet by eighteen A Warehouse twenty feet by fourteen

A Workshop with Blacksmiths forge ... in all about forty two feet long by eighteen ...

A provision Store twenty eight feet by twenty, with loft above. Fourteen Outhouses of various sizes ... Oil house, Smoke house, Wash house for transient persons, and houses for Stock, Vegetables, boats, etc.

Close to the Superintendent's dwelling is the Flag Staff, with Steps leading to the top which is sixty-five feet high, and on the top of which is the lookout, being from one hundred and twenty to one hundred and thirty feet above the level of the sea. The Lookout commands an excellent view of the Island, and the Ocean in nearly all directions. This structure is a good mark for vessels at a distance on visiting the Island and must have cost considerable labor in erecting it.

Moveable Property at Head Quarters [includes]

35 head of horned cattle of which 11 are milch cows
8 Working horses with harness
6 setts New Harness to replace old
6 Riding horses with five saddles—for the use of men going to wrecks and going rounds of the Island in thick weather.
13 pigs
6 geese and forty fowl
1 Life Boat and gear
5 Boats with gear complete
1 Punt
1 large Scow
5 Carts
1 Truck 3 Sleds
2 Harrows 1 Boat Waggon, when separated makes two carts
. . .
1 Bait Mill 3 Boat Sails

A great quantity of shingles made on the Island and about 3000 feet pine Lumber and scantling The Store house contains a great variety including:

10 barrels Pork, sent to the Establishment 1-1/2 barrels Pork, Island cured
3 barrels Beef, from Halifax (bad) to be returned.
Assorted barrels of Herring, Mackerel, Codfish and barrels of Salt.
19 Barrels of Bread [Hardtack]
21 Barrels of Flour
Assorted Barrels of Sugar, Molasses, Oatmeal, Cornmeal, Coffee, Rice, Peas, Barley.
1/2 barrel of Powder [Gunpowder]—saved from frigate Barbadoes wrecked in 1812
1-1/2 kegs ditto
1 bag shot 3 puncheons Soldiers Coats [for the use of wreck survivors]
4 scythes
And draw knives, Axes, glass, whitelead, Salt Petre, Steel yard, Long Composition Plate, and Composition Rudder Gudgeon.

The Oil House contains 2 Large pots and Gear for trying out Oil.
In Small House adjoining:
9 barrs Steel
A lot of Old Copper
A lot of Panel Doors from Barque Patriot
Copper hinges and a Franklin Stove.

    The Workshop contains: Jack Screws, Grindstone, Whip Saws, Crosscut saw, Handsaws, 15 fathoms half inch Chain, Iron Barrs, rods steel, Vice, Anvil, bellows, Forge Hammers, broad Axes, Mauls, Augers, Chopping Axes, Shovels, hammers, Ship's Winch, Grapnel and a large Gin.

    The Warehouse contains 60 sheets New Copper. The Warehouse holds mostly empty Water Casks and 12 Barrels of Bait.

Items listed for the Superintendent's House include:
    Stomach pump and injection pipe, Tooth Drawer, Medicine Chest, a spy Glass that had been on the Island since 1803, 5 Old Muskets, 23 Volumes Books—presented by Bishop Inglis, 14 Life Preservers, 10 rugs, 6 pairs of sheets, 3 pairs Blankets, etc.

The 12-pound cannon or Carronade, in which Mrs. Morris first made her butter was then in its correct position at the foot of the flag staff.

East end of the Lake Station
This station was about 9 miles to the east of Main Station and in a very poor dwelling house lived John Stevens, his wife, child and one other female. There was a barn and stable which would accommodate some 25 head of cattle and held 15 tons of hay. A warehouse measured about 50 by 34 feet. The warehouse, workshop and flagstaff complete the list of structures.

    Townsend recommends that a new dwelling of larger size be built for "should a large number of persons be wrecked in this vicinity in the Winter Season ... there would be much suffering from want of room before they could be sent to Head Quarters."

    Besides many of the items listed for Main Station, Stevens cares for 3 horses and Harness, 3 cows, 3 yearlings, 3 sheep, 6 geese, 12 fowls.

East End Station

This station was about five miles to the east of the East End of the Lake Station and "about four miles from the East Point of the Island at the end of which is dry beach about three miles in length". At this station, in a good dwelling house lived John Nisbett, his wife and six children. There was a warehouse, 40 by 24 feet, a barn and stable accommodating 15 head of cattle and containing five tons of hay. A shed, workshop, outhouses and flagstaff are listed. The usual list of farm, household and marine items are listed as well as 3 horses and their harness, 2 cows, 2 pigs and a good boat and gear.

The West End Station

This station lay about five miles west of Main Station and about one mile from the west end of the island. The dwelling house there measured 20 by 15 feet, the stables, with loft containing hay above. A boat house with good boat was listed. This station was not occupied unless a wreck occurred in the vicinity or work was carried out there. However, the house was "furnished with Matches, Fuel and directions to the Superintendent are hung up facing the door. The dwelling is merely a temporary residence for persons who may be thrown on Shore in the Vicinity, as they would be removed, as soon as discovered to Head quarters."

In 1848 the long central lake, named after Nova Scotia's provincial secretary, Michael Wallace, was continuing to shrink in size. For as early as 1812, the Sable Island Commissioner's report:

> The Pond on Sable which flowed and was navigable for large boats for nearly fifteen miles from West to East in the Island, and which afforded great facility to the transportation of beached property to the place of Shipment, and also for the collection of Fuel, etc., has unfortunately been nearly choaked up by the drifts from the Sand Hills and shifting sands from the Storm of September last [1811] so that all transportations must now be effected by strong horses and teams as the horses produced on the Island, even if tamed, are too small and light to be of any use—at present they are only two horses fit for service and one of them [Jolly] is 18 years old.

It has been said that over-grazing has been the cause of the island's shrinking in size and that the destruction of grass roots allowed the elements to wear away the sand, yet here was a changing landscape despite the fact that the island then supported but 70 head of horses. Lake Wallace was making its exit—not from the over-grazing of the outer dunes, but due to storms and sea damage.

Chapter 14
Strawberries Ungathered

The tall ships no longer crowd on sail nor beat their way across the seas to port and profit and, with rare exception, the "Dark Island" and its surrounding shallows no longer gather in their sorry toll of wreck and life. The lonely, weatherbeaten patrol on his shaggy island horse is seen no more travelling the sandy miles, and the only wrecks discovered today are those whose stark remnants of hull and mast lay newly exposed when the latest storm tears at dune or sea-bed.

Only the wind, or soft whisper of the long grasses, join in chorus with the sea-birds where long ago shrieking, shouting groups of children, released from a week-long sojourn at their dormitory school, mounted their horses and rode hell-bent for home. Not until reaching Main Station, East or West Light, or one or other of the lifeboat stations of later years would this tumbling charge of merry student and half broke horse pull rein. Now none is left to remember. The school house where once Miss Ancient reigned supreme is gone, lying in ruins and open to the storms stand the once busy stations—the creeping sands are slowly claiming all.

The strawberries lay ungathered, the cranberries unharvested, and where the wide felloes of Boutilier's wagons left wide wakes of crushed fruit, some privileged student will be searching out the island's varied fauna or flora. To study there is an honour bestowed on these young people by governments well aware of the fragile nature of Sable Island's fabrique; delicate, and yet, much of it surviving the buffeting of the ocean elements. It is a world that would soon disintegrate, as has so much of the east coast of the United States of America, were large numbers of visitors and vehicles to travel the island.

Only a handful of people—the staff of the weather station—live on Sable Island now. Years ago, forty or more would gather to greet the periodic arrival of the supply ship out of Halifax. When the author landed there in April 1980, the little plane dropped out of the clouds and everything on wheels came racing along the sands. As the visitors touched down on what was once the long reach of Lake Wallace, a delightful group of islanders and their bouncing dogs gathered round for news, supplies and a change of faces for a while. Everyone piled into truck or trailer for the long ride back over the beaches with their sea-tossed litter of sand etched bottles, plastic buoys and torn fishing nets. Horses lifted their heads for a moment to watch and herds of seals made for the sea in disconnected undulating humps.

The station took on the bustle of Christmas morning when mail and ordered supplies were sorted out. Evidence of a changing order for the island was all around. A rigger, flown in with the supplies, went off to repair a pylon; tractors with wide tires kicked up the sand as they covered in minutes distances that would take a team of horses all day. Fuel drums stood nearby, and in the garage, where the visitor made conversation with some disembodied voice beneath a failing truck, for which "they still haven't sent out that spare part," the talk turned, inevitably, to Sable Island's oil wells and off-shore drilling rigs.

Fifteen years have brought many changes in Canada's economy and, in supplying a constant need for oil, rigs now function offshore but Sable Island itself has been spared their erection.

It would be more than regrettable if in new shifts of policy and reorganization in the 1990s the unique and historic world of the island should be badly managed or, worse, abandoned. Without wise planning and control, this island world, so tied to Canada's early history, development and commerce, could quickly revert to conditions existing before Nova Scotia took active control in 1801. Before that date and the establishment of lighthouses and lifesaving stations, its fate was one of constant pillage, slaughter and destruction.

Too much of the world today has been destroyed by man's ambitions and lack of thought. Wise, informed heads must now work and plan to prevent such a reversal, for, as to the horses alone, they have no world equal.

The mind focuses back to a day on the island when a tough young stallion, his lone mare and her promising youngster ranged in the vicinity of the weather station, three of a very few island horses in easy contact with humans. With his family behind him, he approached newcomers warily, but lean and ragged-hipped before the arrival of spring grass, he took with surprising gentleness, a frozen brussels sprout from the hand of a small child.

The stud herded his group back up the hill. He had correctly assessed the quantity and quality of any possible handout, then the security of distance took precedence over hunger. It had been one of the periodic difficult winters and the mortality among the horses was higher before the island showed green again. Unlike the British ponies of Dartmoor, Exmoor, the New Forest and the Welsh hill, and elsewhere, who also live out at all seasons, these Sable island horses, now living entirely independent of man, have no supporting hay during poor weather.

This was always the fate of the animals of early mainland settlers. With winter feed all but nonexistent, disease plagued many, and death claimed others as they lived out in forest and thicket until spring. Even animals kept in barns became weakened through poor, or scarce feed, and at the beginning of each day many had to be lifted to their feet. This poor winter grazing is not foreign to the Sable Island horses, and to feed them—that is, if they will eat hay air-lifted to them—means bringing along weak stock, and for this breed of horse, a greatly increased birth rate and numbers of horses which the island cannot support.

The herd management which once ordered the horses' lives is no more; the annual drives no longer take place. The skills employed by the islanders in driving and corralling have been lost and attempts at shipping off any number of them today would present problems. The last shipment arrived at Hantsport, Nova Scotia, quite a few years ago and a number of horses were injured; one had to be destroyed. Moving large numbers off the island required suitable vessels, seamanship, a knowledge of the sea about the island and, last, but far from the least important, a working knowledge of the horses themselves.

What if it were possible to ship them off without mishap? What would happen to them on the mainland? The old markets for them no longer exist, and for most, the only place for them in the modern show ring or horse event is as a novelty. Out crosses of him would see the end of his breed, or type; as a pet he would be a five-minute wonder, then he would find himself sold on to increasingly poor owners and tasks until he finally reached the butcher. He could, of course, serve his time in the poorer class of hacking stable or trekking establishment, but his situation there is little better and his work load greater.

What are his chances in a world that increasingly sees sound young animals sold as slaughter horses; the meat marketed for human consumption? A number of British pony breeds suffer degrees of this fate and an owner can often get a better price for his stock from the exporter who sells them for meat, than from those who purchase saddle or light draught animals. Newfoundland is making valiant efforts to save their native horses from the same fate, for continuous shipping off, would see that old breed extinct. In the rest of Canada and the U.S.A., along with worn out and cripples, many good horses are going to the slaughterer. The total numbers for each country are probably unknown. Is this to be the fate of these unique Sable Island horses?

Ideas as to their fate when they periodically become of wide national interest are varied, and range from those who, not realizing that many of the Sable horses can be very rough customers, kindly offer one or two a home. Others would have them shipped off the island and halfway across the country before they would be used for pet food or mink feed.

One suggestion was to send out 'sharpshooters', who would gun down the lot; one bright, enterprising soul wished to establish fox farms there and shoot the horses as required for feed. How would a Sable Island dialogue run then?—*What do you want today, Bill, a sixty pound foal or an eight hundred pound colt? Bay or chestnut? The last boat load of tourists cleaned us out of hides and sparrows eggs!*—Such thoughts can be extended to a point approaching idiocy and yet, regrettably, still not go beyond the limits of the possible, should wise heads and legislation not prevail, as to the island's future, or the average Canadian not respect his history and natural resources.

What then, the reader might ask, is to become of these horses? Within the context of a Nature Reserve, nothing need happen to them at all. It will be said that the island is slowly wearing away; but in one area it is at the moment adding five miles to its length. The horses are hungry in the winter. Studies show that the horses die with food all around them. It is, however, food of poor nutritive value at that time of year. With wet, cold weather, the systems of some are unable to process such food, or obtain sufficient heat and sustenance from it. Unable to get to their feet, as was the case with mainland horses in early days, they simply weaken and die. Better for a horse to die with the sky above and a familiar world about him than to stand close-tied and half forgotten as do many stabled animals today. Such animals are not only starved from lack of food, they often suffer loneliness and poor shelter. Old hands, used to the ways of Sable over many years, will tell you that although the horses graze heavily during the winter, they will, when the grass grows long, only eat some of the grass from each tuft, unlike sheep or cows, who will crop close.

What about keeping some in parks, etc.? What has been the fate of another rare horse, the Przewalski? The following is from *International Wildlife* (November-December 1977).

A study of the horse by Jon Bouman of Rotterdam, shows that the fertility rate is decreasing, the number of abortions is up and the life expectancy is decreasing. Congenital defects, including stunted growth and hip weaknesses in mares, are on the increase. In addition, the relatively small size of the animal's pens (compared to their large range in the wild) has begun to modify behaviour and breeding; so has the artificial size of the captive herds ... Bouman believes zoos could use their stud books to breed animals more efficiently. He has called for quick establishment of large reserves for free-roaming herds. Failure to reestablish relatively natural populations of the horse within ten years ... could mean that the species will by then be so weakened that wild herds will be impossible.

Bouman's observations are sound, but, in the case of the Sable Island horses, were the process of natural selection removed and their habitat changed, the first would rob them of part of their scientific value, the second would produce animals of altered conformation.

Those people only interested in the pure Barb feral horse may have difficulty coming to terms with the fact that the herd has seen a number of introduced stallions. It should be remembered that these Sable Island horses were a mixed breed for years before arriving at the island—that is, if their Acadian origin is agreed upon. Those looking for the pure mustang, may fly over feral horses on our, and American plains, and leave untouched those which they observe to enjoy five lumbar vertebrae (Barb characteristic?), but gunning down for pet food those which labour under six vertebrae. Can this distinction be made among Sable Island horses?

Dr. D. Welsh carried out considerable morphologic studies on these lines on the island horses. His findings read:

> The Sable Island horses are highly variable in character, and traits used to separate types of horses frequently co-occur in the same animal. Examination of the spinal column showed that thoracic, lumbar, sacral and caudal vertebrae number vary, as do several characters within these segments. Comparison with published data indicates that the Sable Island horses are as variable as *Equus caballus przewalskii*, normally considered the most variable of the equids. This high variability throws doubt on the common practice of classifying horses by the number of lumbar vertebrae.

Despite the fact of introduced stallions, these horses have bred by a process of natural selection for close to 250 years— probably longer, as the French did not geld their horses and, in Acadia, they ranged freely—thus, one of their chief values lies in their genes.

It is increasingly maintained that the domestic breeds of horses are 'all falling apart'; one or the other of Navicular disease, Ataxia and founder are to be encountered in fashionable lines of horses. Geneticists, along with conservationists in other fields, are concerned; the genetic base is becoming so narrow that "only under wild, or semi-wild conditions, where mares have a choice of stallions, will horses remain viable." One adds, "Even fruit flies will refrain from in-breeding when given a chance."

Along with seed, sperm and other banks, now being assembled to assure healthy continuance of our plants and animals, the genes in the Sable Island horse should be valued and conserved. Their living influence may yet serve as a stabilising element in an off-balance world of horses.

Man has interfered with the genetic pattern of so many horses in breeding for fashion, speed or specific function; we now must begin preparations to right some of these wrongs.

Postscript
by Zoe Lucas

Sable Island is an isolated crescent-shaped sandbar surrounded by the cool waters of the western North Atlantic Ocean off the east coast of Canada. The island is located at 44°N, 60°W, about 160 km from the nearest landfall in Nova Scotia, and approximately 300 km south east of Halifax. It is roughly 42 km long and 1.5 km wide and has an area of about 3,425 ha.

The shape of Sable Island is the result of atmospheric and oceanic influences. The shape and position of the dunes reflect the prevailing westerly wind direction and storm trends (Hennigar 1976). From day to day, month to month, ocean currents, waves, and tides modify the width and contour of the beaches and change the dimensions of east and west spits. The length of the spits at both the east and west ends of the island wax and wane as sand is moved back and forth along the beach by waves and currents. Along the north and south sides of Sable Island are many areas where the high dunes can be seen collapsing into the sea, however, in other parts of the island colonizing vegetation is building new dunes and stabilizing previously naked sand.

In summer and autumn much of this sandbar is cloaked with lush, green vegetation and wildflower blooms; in winter and early spring its dunes are rather bleak, grey and windswept, and appear deceptively devoid of vegetation. Except for one small decumbant pine (about 45 cm in height) surviving from a planting near the weather station over twenty years ago, there are no trees on Sable Island. Studies of pollen and spores on the island indicate that the area has been treeless for the last 11,000 years or more.

Now about 40 percent of the island's land surface area of 3,425 hectares is covered by vegetation, comprising over 175 plant species in several distinctive plant communities (Catling et al. 1985). These include the colonies of sandwort on the east and west tips of the island; the shrub-heath and cranberry communities dominated by juniper, crowberry, bayberry, wild rose, blueberry and cranberry; and the richly vegetated freshwater pond and pond edge communities. The low undulating heathlands are the oldest plant communities on the island, and are found in inland areas sheltered by the north and south dunes. In wetter areas of the heath there are cranberry hollows, often rich with violets and orchids—seven species of terrestrial orchid, including the grass-pink, survive and bloom in the island's heaths and bogs. Generally the freshwater ponds are bordered by plants such as iris, violet, buttercup, swamp candle, and orchid, with stands of bulrush and spike-rush. Knotweed, burweed and water milfoil are aquatic plants found in many of the ponds. The grasslands are by far the largest community (comprising roughly 30 percent of land surface) and are dominated by beach grass with several other species in varying abundance, including beach pea, bluegrass, seaside goldenrod and yarrow.

Sable Island is home and sanctuary for a variety of animal life. Of the more than 600 invertebrates found on the island, three moths, a beetle, a nematode and a freshwater sponge are endemic fauna (Wright 1989). More than 300 species of birds have been sighted on Sable Island. The island is home to a number of breeding species and offers habitat for migrating shorebirds and waterfowl. The Ipswich Sparrow—a large and pale subspecies of the Savannah Sparrow—nests exclusively on the island (Stobo 1975). Nesting colonies of Arctic and Common Terns, and several pairs of the rare Roseate Tern—threatened and declining throughout most of their range—are found on the dunes and beaches. More common avian nesting residents include Herring and Great Black-backed gulls, several species of shorebird and five species of duck. Migrating shorebird visitors include Whimbrels, Willets, Sanderlings, White-rumped and Semipalmated Sandpipers, and Ruddy Turnstones (McLaren 1981).

Two species of seal are year-round residents on the beaches and in surrounding waters of Sable Island. Populations of Grey and Harbour seals, during January-February and May-June, respectively, give birth, suckle their young and breed on the island. The Greys, with the males weighing as much as 270 kg, are the larger and more numerous of the two species.

The most famous of the island's fauna are the feral horses. Aside from the few human inhabitants, the Sable Island horses are now the only terrestrial mammals on the island. Although access to the island is restricted, by both location and regulations, the horses of Sable Island are well-known. Because they are an introduced species on the island, they do have detractors. There is some concern about their impact on the island's vegetated terrain, nevertheless the horses are valued and appreciated by those having an interest in history, in biology and behaviour of equids, and in conservation of minor breeds and genetic resources. During the early 1970s the Sable Island horses were the subject of study for the doctorate work of Daniel Welsh (1975), and since then researchers have examined various aspects of the horse population on the island (e.g. Lucas et al. 1991). Indeed, there has been international interest—behavioural research by biologists from the Equine Museum of Japan culminated in an exhibition, in Yokohama in 1994, devoted to the life of the Sable Island horses. In addition, the horses have been featured in several nature films and numerous books and magazine articles produced both in Canada and abroad.

The small sturdy Sable Island horses appear to be well adapted to life on this isolated sandbar, living in an environment not at all as harsh or as bleak as portrayed by some writers. The island's climate is temperate oceanic and is generally milder than that of mainland Nova Scotia. Winter temperatures are normally between +5° and -10° Celsius; summer temperatures peak in August at 25° Celsius. Though the Sable Island horses graze many of the 175 or more plant species on the island—including the eelgrass of saltwater Lake Wallace—the primary forages for the horses are beach grass (*Ammophila breviligulata*) available year-round, and beach pea (*Lathyrus maritimus*) and sandwort (*Honckenya peploides*) available from spring through autumn.

During summer and autumn the horses graze in the lush vegetation on the high dunes and the sheltered inland fields, gaining weight and storing energy for the coming winter. In October and November the horses grow thick, woolly coats. During winter they gradually lose weight, and by the time spring arrives have become quite thin. In late May and June, though they look very scruffy as the winter coat is shed in matted rags, their body condition improves as fresh summer growth of beach grass, beach pea and bluegrass covers the dunes. Thus the horses experience an annual cycle of fat and thin, a pattern which suggests an overall balanced energetic budget with a positive and negative phase. It is popular wisdom that horses can thrive in habitats where cattle do poorly. Ruminants, such as cattle and sheep, are limited in the amount they can eat by the rate at which their multi-chambered stomachs can process food. Non-ruminants such as horses are not limited in this way, and one reason for the success of Sable Island horses may be that with the reduced quality of available forage in winter, they can eat and digest greater quantities in order to compensate for poor quality.

Drinking water is available at the numerous freshwater ponds on the island. In addition, the horses dig waterholes in hollows and low areas between the dunes, and keep the holes open year round except in the very coldest weather. Severe winter cold is rarely prolonged, but when ponds and waterholes are temporarily frozen, the horses will eat snow. During days when heavy snowfall covers the ground the horses are quite capable of grazing. They use their hooves to 'paw' away the snow or with their noses push the snow off the vegetation as they move along and eat. However, with the island's mild winter temperatures and high winds, snow accumulation seldom persists. Even in winter the island environment can be rather benign and certainly not as harsh as the mainland. Though there are no trees on the island, shelter from the wind is available in the lee of beach dunes (some of which are as high as 25 meters) and lower heathland hills.

The Sable Island horse population usually comprises 40 to 50 family bands. Band structure is variable, but most often consists of one dominant stallion, one or more mares and their offspring, and occasionally one or two subordinate males. Average band size is usually four to eight, though bands of ten and twelve individuals have been observed. Males who are not in family bands, form loosely organized "bachelor" groups, or, particularly if they are older, live as solitary stallions.

The last three decades of aerial and ground surveys show that since the 1960s the horse population has fluctuated between 165 and 450 individuals. In the absence of human interference and management, natural selection occurs as the hardier individuals survive to pass on to their offspring characteristics that make them suited to life on this offshore island. Mortality occurs mostly during late winter and early spring, and is strongly correlated with certain weather conditions. Unusually cold and wet winters result in greatly increased mortality, most often affecting the very young and very old horses, though age-related mortality patterns are variable. However, in mild winters the mortality rate can be 5 percent or lower.

Surveys carried out by Canadian Coast Guard between 1960 and 1980 were aerial counts made using a fixed-wing aircraft or helicopter. While such surveys provided accurate data on the number of horses, they did not provide more detailed information about body condition, foaling rates, sex ratio and age structure. Since the mid-1980s a ground survey of all horses on the island has been carried out at least once every ten to twelve weeks. As part of the long-term program of population monitoring, the location of family bands and bachelor groups is routinely checked, individual condition recorded and birth dates for foals are determined. Whenever a horse is found to be missing from its usual group, a search of the area is made to determine if the horse has moved into another group, is solitary or has died. All carcasses are examined for obvious injuries, and, in the case of mares, for pregnancy, and when possible a necropsy is conducted. The carcass is marked for collection of skeletal materials later in the season. The bones are used for a number of studies (e.g. examination of tooth wear), and eventually will be deposited in a museum collection.

Many observers have used the modern domestic horse as a standard when describing the Sable Island horses. Thus, comments such as "head large and ill-set on", "lacking in quality and characterized by a fairly large, plain head", "droopy quarters" or "poor hindquarters", "narrow chest and knock- knees", and "cow-hocks" are common. However, Sable Island horses are a product of both their lineage and their demanding environment. The population is composed of survivors—horses that are able to thrive while foraging for themselves, digging for water, finding shelter, galloping up and down high dunes and along sandy beaches, exposed to storms and high winds, giving birth unassisted, and living without feed supplements or anthelmintics. Modern domestic horses, however, are the result of long-term selective breeding "for fashion, speed or specific function", and while they may adequately fulfill their assigned role in man's scheme —in harness, dressage, show jumping—they are not necessarily fit. Numerous domestic breeds of cattle, dogs, pigs and poultry have also become less fit genetically as a consequence of intensive selection for qualities that serve rather limited purposes of humankind. With increasing concerns about the deterioration of domestic breeds of horses, the Sable Island population are an important genetic resource. In some cases the feral and free-ranging counterparts of domestic animals are actually reservoirs of healthy genetic material, and are often the only source of information about the pre-domestication lifestyles of animals that no longer exist in the wild. However, it is also true that feral and free-ranging animals may have a negative impact on habitats and native flora and fauna in areas where they have been deliberately or accidentally introduced by people. Feral plant-eating mammals have caused the loss of endemic plant species and environmental damage in such well-known island habitats as Hawaii, the Galapagos Islands, and Santa Catalina Island (California). Understandably such well-documentated negative

A family band drinking at a waterhole on the beach at the edge of the dunes — the waterholes, which the horses dig with their front hooves, can be as much as 45 cm in depth during August.(Photo Zoe Lucas)

A stallion and a mare eating beach grass — in July, August and September the horses gain weight as they graze in stands of tall green beach grass which, in some areas, is more than 100 cm in height.(Photo Zoe Lucas)

A family band grazing—in summer and autumn the vegetated dunes of Sable Island provide lush pastures of beach grass and thick, tangled patches of beach pea, bluegrass and fescue.(Photo Zoe Lucas)

Three bachelor males eating aquatic vegetation—during summer horses living on the western third of Sable Island spend many long hours wading in the freshwater ponds to graze. (Photo Zoe Lucas)

impacts raise concerns about the effect of feral horse activities on Sable Island's native plants, birds and invertebrates.

There have been occasional calls for management programs by individuals and groups who are concerned either about the welfare of the horses, or about the impact of the horses on the island and its native species of plants, birds and invertebrates. In most cases the arguments are flawed by lack of information, and frequently overlook the importance of this population of feral horses. The Sable Island horses differ from other feral groups in several ways. Most other groups of free-ranging or feral horses are subjected to some kind of management (e.g. culling or fertility control) or are exposed to occasional influxes of domestic stock. Because they are isolated and have been free of human interference for over thirty years, the Sable Island horses are of great scientific value. A series of long-term research programs have been under way since 1982. These include studies of equine parasites, fertility and reproduction, forage quality and grazing behaviour, and population monitoring programs. All current research on the Sable Island horses is non-invasive. The horses are protected by law from interference by people, thus research data are collected by observation and from examination of waste materials or tissues taken from horses that have died of natural causes.

Though the horses are protected by law, that protection is in effect only so long as the horse population and their island habitat are monitored. Since the first days of the life-saving establishment in 1801 there has been a continuous government presence on the island, and also since then, access to Sable Island has been controlled by the government. Though Canadian Coast Guard has responsibilty for protection of the Sable Island horses, the full-time federal presence on the island is now embodied in the Environment Canada weather station. Weather data has been collected continuously on Sable Island since 1891 to the present, but during the last few decades the station has come to serve a far broader role on the island. Besides having an important role in global atmospheric research and monitoring, the station provides an infrastructure which enables a wide range of other activities —including research on horses, seals and birds, vegetation and sand transport, and ocean pollution monitoring. The weather station is the hub of operations on the island, providing support in landing aircraft, communications and safety, and its presence ensures that regulations and guidelines regarding the island terrain, flora and fauna are respected.

Barbara Christie has provided us with a fascinating examination of the history of the Sable Island horses, and yet there is still so much to learn about the behaviour and biology of the present-day Sable horse—so much to learn about the complex relationships between the horses and their unique Sable Island habitat.

References

Catling, P.M., B. Freedman and Z. Lucas. 1985. The vegetation and phytogeography of Sable Island, Nova Scotia. Proceedings of the Nova Scotian Institute of Science.

Hennigar, T.W. 1976. Water resources and environmental geology of Sable Island, Nova Scotia. Report No. 76-1. Nova Scotia Department of the Environment, Halifax, N.S. 56 pp.

Lucas, Z., J.I. Raeside and K.J. Betteridge. 1991. Non-invasive assessment of the incidences of pregnancy and pregnancy loss in the feral horses of Sable Island. J.Reprod. Fert., Suppl. 44:479-488.

McLaren, I.A. 1981. The birds of Sable Island. Proceedings of the Nova Scotia Institute of Science 31:1-84.

Stobo, W.T. and I.A. McLaren. 1975. The Ipswich Sparrow. Proceedings of the Nova Scotia Institute of Science, Vol. 27, Suppl. 2, 105 pp.

Welsh, D.A. 1975. Population, behavioural and grazing ecology of the horses of Sable Island. PhD Thesis, Dalhousie University, Halifax N.S. 405 pp.

Wright, B. 1989. The fauna of Sable Island. Curatorial Report Number 68, Nova Scotia Museum, Halifax Nova Scotia.

GLOSSARY

Andalusian: The Iberian (Spanish horse crossed with Barb became known as the Andalusian, taking its name from the province in the south of Spain, Andalusia.

Barb: Originated in Morocco and Algeria, and stood 14-15 h.h. The characteristics were flat shoulders, rounded chest, relatively long head, and tail set lower than an Arabian's. The hair of the main and tail were profuse. It lacked the spirit of the Arabian and was of less refinement, but was of hardy constitution and docile temperament. Owing to cross breeding, the purebred Barb is now hard to find.

Brood mare: A mare used for breeding

Covering: The act of generation between a stallion and a mare

Dished face: Concave

Entire: Ungelded Horse

Feral: Having escaped from domestication and become wild, not domesticated.

Gelded: Emasculated

Hand: h.h. The measurement by which the height of a horse is reckoned; it equals four inches.

High School airs: Classical art of riding as practised today by the Imperial Spanish Riding School of Vienna with the Lipizzaner breed of horse.

Mustang: A term applied primarily to the feral and semi-feral horses of American and Canadian western plains. Probably of Spanish origin, originating from earlier Spanish imports of Spanish-Arabian and Barb blood. Seldom more than 14.2 h.h., it was scraggy and tough, hardy, courageous and possessed a cast iron constitution.

Purebred: Pure of its type

Put down: The destruction of a horse for reasons of health, age, disposition, etc. On Sable Island, it was done by rifle bullet.

Roman nose: Convex

Stallion: A horse, not under four years, capable of reproducing the species.

Standardbred: The official name of the famous American trotting and pacing horses, bred with extreme care and scientific efficiency. The father of the breed was Rysdyk's Hambletonian, who descended from the American imported English Thoroughbred, Messenger.

Standing: The nature of the flooring on which a stabled horse stands.

Standing: To have a stallion at a certain location available for breeding purposes

Thoroughbred: Synonymous with present day race horse and best known of all English breeds. All thoroughbreds trace their ancestry to three sires, the Darley Arabian, Godolphin Barb and Byerley Turk.

a, Muzzle; *b*, Gullet; *c*, Crest; *d*, Withers; *e*, Chest; *f*, Loins; *gg*, Girth; *h*, Hip or ilium; *i*, Croup; *k*, Haunch or quarters; *l*, Thigh; *m*, Hock; *n*, Shank or cannon; *o*, Fetlock; *p*, Pastern; *q*, Shoulder-bone or scapula; *r*, Elbow; *s*, Fore thigh or arm; *t*, Knee; *u*, Coronet; *v*, Hoof; *w*, Point of hock; *x*, Hamstring; *zz*, Height.

BIBLIOGRAPHY

A complete bibliography would be disproportionate to the length of the book. The following are some of the main works consulted.

Boutilier, Superintendent R. J., Sable Island letter books, diaries, originals 1884-1911. PANS Vols. 4, 5, 6, 7.

Cathcart, Earl, "The French Government's Method of Improving horses" Journal of the Royal Agricultural Society of England, Vol. 25, (Oct. 1889).

Campbell, L. G., History of Sable Island Before Confederation, M. A. Thesis, Dalhousie University, 1962.

_____, Sable Island, Fatal and Fertile Crescent, Lancelot Press Windsor, N.S.

Clark, Andrew Hill, Acadia, the Geography of Early Nova Scotia to 1760, University of Wisconsin Press, Madison, Milwaukee and London 1968.

Colonial Office Papers /217/76, 77, 79, 1801-1804. Journals and reports of superlntendent James Morris

Dent, Anthony, and Goodall, Daphne Machin, The Foals of Epona, Gallery Press, London, 1962.

Douville, Raymond, and Casanova, Jacques Donat, Daily Life in Early Canada from Champlain to Montcalm, Translation by Carola Congreve 1st American Ed., Macmillan, New York, 1968.

Erskine, J. S., The Ecology of Sable Island Wolfville, N.S., 1952.

Fowke, Vernon C., Canadian Agricultural Policy, The Historical Pattern, University of Toronto Press, 1946.

Gagnon, F. E., Le Fort et le Chateau Saint Louis, 1896,

Ganong, W.F., The Cadillac Memoir of 1692, 1930

Gilpin, J. Bernard, "On Introduced Species of Nova Scotia", Transactions of the Nova Scotia Institute of Natural Sciences i, (printed ii) part 2,

_____, Sable Island: Its past history, present appearance, natural history, etc., Halifax: printed at the Wesleyan Conf. Steam Press,

Goodall, Daphne Machin, A History of Horse Breeding, Robert Hale, London, 1977.

Halleck, Charles, "The Secrets of a Sable Island", Harper's Magazine, vol. 34, 1866.

Hanny, J., The History of Acadia from Its First Discovery to Its Surrender to England, J. & A. McMillan, St. John, N.B., 1879.

Howe, Joseph, "Condition and past management of the humane establishment at Sable Island" A report to Sir John Harvey, Appendix to Journal of House of Assembly, Province of Nova Scotia, No. 24 (1851) and No. 8 (1852).

Journals of the House of Assembly of Nova Scotia (covering a period

1801-1867)

Kalm, Peter, Travels Into North America 1748-1761, Sweden. Translation into English by John Reinhold Forster and son, 1770.

Lanctot, G., L'establissement du Marquis de la Roche a LIle de Sable, Canadian History Association, Ann. Rep., (In French) 1933,

Le Mercier, Rev. Andrew, "The Island Sables [Notice published in the Boston Weekly News Letter, February 8, 1753, and quoted in St. John's Article on Sable Island]

L'escarbot, Marc, The History of New France by Mare L'escarbot, with English translation, notes and appendices by W. L. Grant and introduction, H. P. Biggar, Champlain Society, Toronto, 1907-1914.

MacDonald, S. D., Art. III: "Notes on Sable", Proceedings and Transactions Nova Scotia Institute of Natural Science, Vol. 6, 1883.

_____, Art. V: "Sable Island" (cont'd) Proceedings and Transactions Nova Scotia Instituteof Natural Science, Vol. 6, 1884.

Martin, A., "Extracts from the Journal of the late Captain Farquhar: his stay on Sable Island", Collection of the Nova Scotia Historical Society, No. 27, 1946.

McLaren, I. A., "Sable Island, our heritage and responsibility", Canadian Geographic Journal, 1972.

Moorsom, Capt. W., Letters from Nova Scotia, Henry Colburn and Richard Bently, London, 1830.

Patterson, Rev. George, "Supplementary Notes on Sable Island", TRSC, 2nd ser., III (1897), Section II, 131-8.

_____, "Sable Island: Its History and Phenomena", TRSC, 2nd ser., XII (1894), Section II. Published separately in Montreal: W. Drysdale & Co., Halifax and Pictou, 1894.

Perkins, Simeon, The diaries of Simeon Perkins 1766-1780, H. A. Innes, Ed., Champlain Society Publication.

Public Archives of Nova Scotia Manuscript Documents. Papers relating to the government establishment at Sable Island. 424 vol. 1 1801-1820; 425 vol. 2 1809-1865; 426 vol. 3 1825-1847; 426-1/2 vol. 4 1852-1858

_____, Log books of schooner Daring running between Sable Island and Halifax 427 vol. 1 June 1853—Oct. 1856; 428 vol. 2 Oct. 1856—Nov. 1858; 429 vol. 3 July 1859—Aug. 1862

Robinson, John, and Rispin, Thomas, Journey Through Nova Scotia, Containing A Particular Account of the Country and Its Inhabitants, Printed by the authors by C. Etherington, 1774.

Saunders, Richard M., "The Introduction of Plants and Animals into Canada", Canadian Historical Review, Vol.16,1935.

Séguin, Robert-Lionel, La Civilisation traditionelle de "l'habitant" aux 17e et 18e siecles, Montreal, Fides (1967)

St. John, Harold, Sable Island, with a Catalogue of its Vascular Plants, Procedings of the Boston Society of Natural History, Vol. XXXVI, No.1,1-103, with plates 1 and 2. Published separately Boston: Printed for the Society, 1921.

Welsh, Daniel E., Population, Behavioral and Grazing Ecology of the Horses of Sable Island, Nova Scotia, 1975.

———, "The Life of Sable Island's Wild Horses", Nature Canada, April-June, 1973.

Wentworth, Sir John, Observations upon an establishment proposed to be made on the Island for the relief of the distressed and preservation of property with a statement of facts. Canada Archives Report 1895.

Young, Arthur, Travels in France during the years 1787-1788, Miss Betham-Edwards, Ed., Bell and sons, London, 1900.

About the Author

Born in Sussex, England, Barbara Christie came to Canada after her war-time marriage to her Nova Scotian husband. Of a family long interested in horses, she continued to participate in equine affairs in Canada.

With her family grown, she combined a love of history and horses and began a 22-year study of horses in the Atlantic Provinces, early New England States and Europe. It is a comprehensive work covering a period that runs from the first French horses to arrive at Port Royal in 1611, to roughly the year 1900. Listed in the standing research are hundreds of stallions of a number of breeds and types, many imported along with brood mares from Great Britain and the United States. Also covered in her research are flat racing, pony racing, hurdles, polo, ice racing and driving clubs.

It is from this broad background study that the history of the Sable Island horse has been taken. It is the hope of the author that those interested in horses and conservation will, through this small book, become better acquainted with these unique and historic equines.

Besides her interest in the world of horses, Barbara Christie is a Research Associate of the Nova Scotia Museum, an enthusiastic gardener and a qualified flower show judge.